AQUACULTURE
MANAGEMENT

AQUACULTURE MANAGEMENT

James W. Meade

An Book
Published by Van Nostrand Reinhold
New York

An AVI Book
(AVI is an imprint of Van Nostrand Reinhold)
Copyright © 1989 by Van Nostrand Reinhold
Library of Congress Catalog Card Number 88–27997
ISBN 0-442-20570-8

Printed in the United States of America

Van Nostrand Reinhold
115 Fifth Avenue
New York, New York 10003

Van Nostrand Reinhold International Company Limited
11 New Fetter Lane
London EC4P 4EE, England

Van Nostrand Reinhold
480 La Trobe Street
Melbourne, Victoria 3000, Australia

Nelson Canada
1120 Birchmont Road
Scarborough, Ontario
Canada M1K 5G4

16 15 14 13 12 11 10 9 8 7 6 5 4 3 2 1

Library of Congress Cataloging-in-Publication Data

Meade, James W.
    Aquaculture management.

    Includes index.
    1. Aquaculture. I. Title.
SH135.M43   1989   639'.0916   88–27997
ISBN 0-442-20570-8

To my parents,
Jim (Wink) and Veda May
for the debt unowed I can never repay

Preface

Although some nations, such as Japan, have invested in aquaculture research and developed major aquaculture industries, the opportunities for similar development in the United States remain largely unnoticed. In a typical recent year the United States, which claims 20% of the world's marine fisheries resources, imported seafood worth $4.8 billion and exported $1.3 billion. In addition to the $3.5 billion deficit in food-fish, was another $2.7 billion deficit for nonedible fishery products. Next to oil, fishery products constituted the second highest drain on the United States balance of payments and accounts for a significant portion of the foreign trade deficit. Furthermore, fish consumption has been increasing in North America. In response to the demand for fishery products, aquaculture managers not only have the opportunity to realize economic profit, but in doing so can make an important contribution to reducing the national debt, providing employment, and enhancing our diet.

This book might be considered a farm management text for those in aquaculture. It is intended to provide an introduction to aquaculture principles and an introduction to management, including business and people management, microeconomics, and the concepts of efficiency and productivity. I hope it will bridge the gap between conservationists, the academic community, and commercial culturists. Abundant references should enable the reader to quickly access literature on most topics germane to the management of culture systems. Many scientific names have been omitted, but they can be easily found by cross reference in such publications as *Common and Scientific Names of Fishes*, American Fisheries Society Special Publication No. 12, 1980 (Bethesda, MD, 174 pp.). Although this is a general or "principles" text, it stresses the component nature of culture systems, each composed of several enterprises that can themselves be studied, managed, and mixed to increase productivity and to maximize returns to the manager's available resources. The concepts are suitable for application by government employees and managers, for university and other not-for-profit organization professionals, as well as for those in the "business end" of the industry.

The book is not intended to be a culture techniques manual. The set of culture techniques for red drum, for various algae, for striped bass, oysters, Atlantic salmon, eels, inland trout, abalone, sturgeon, crawfish, mullet, ornamental fishes, lobster, and many others are each topics of books or book-sized manuals. And new techniques and manuals are appearing at an increasingly rapid rate. Not only do the generalized, and currently changing, culture techniques differ among species-related culture systems, but the techniques may be modified for geographic location, specific site features, genetic make-up of stock and enterprise objectives. Although this book identifies sources of information that address the techniques used in various types of culture, it primarily concerns management concepts as applied to aquatic culture in the United States.

Acknowledgments

I am indebted to the United States Fish and Wildlife Service (USFWS) and Garland B. Pardue, for providing me the opportunity to write this book. Other USFWS personnel were also instrumental. William F. Krise was a principal author of Chapters 1 and 6. Joyce Mann provided administrative planning advice. Susan Bencus not only typed and retyped but provided important technical advice, and Betsy Driebelbies provided technical reprints and library assistance. Lori Redell, Debbie Enderle, and Les Mengel produced the figures. Paul Eschmeyer provided invaluable review and editorial comments. George Ketola provided information on ethics.

    I am grateful to many other friends and associates for their help. Thomas Young of Mansfield University was principal author of Chapter 7. Sandra Linck and Gale Largey, both also of Mansfield University, gave me direction and advice on "people" issues. Carol Fryday, independent editor of Pittsburgh, PA, and Dick Soderberg of Mansfield University, made important editorial and review comments. I referred to lecture notes from Howard Clontz and Sidney Bell of Auburn University, and Robert S. Pomeroy of Clemson University, provided information and advice. Harry Westers of the Michigan Department of Natural Resources, and David S. Liao of the Marine Resources Research Institute, S.C., also provided technical information.

Contents

Part One

PERSPECTIVES

Chapter 1

Principles of Fish Culture and Aquaculture Systems

FOCUS: Perspective on the field of fish culture and aquaculture

HIGHLIGHTS:
- Reasons for culture
- Need for more management
- Types, locations, and organisms of culture
- Temperature effects in culture
- Intensification gradient

ORGANIZATION: Definition, Principles, and Need for Management
The Culture System Spectra
    World Overview
    Fresh Water
    Salt Water
    Temperature
    Organizational Systems
    Operational Systems and Degree of Intensification
Summary

DEFINITION, PRINCIPLES, AND NEED FOR MANAGEMENT

The days of the hunter-gatherer are largely past. It costs less energy—requires less effort—to grow crops—or turkeys, or cattle, or flowers, or mushrooms, or just about anything organic—than it does to hunt and gather those things. Almost anything, that is, except ocean fish. The world's fishing industry is very much alive, but it has approached its limit in harvesting many of the traditional food fishes.

Like other agricultural commodities, fish are reared, rather than hunted and gathered, to save energy. Motives vary from making a profit to ameliorating environmental concerns to providing recreation (among others), but generally the effort is an energy cost saving. Important exceptions include rearing endangered or threatened species, producing specialty products that do not normally occur in nature, and providing a vehicle for education or for social welfare.

Fish culture is the husbandry practice of rearing finfish. It existed in China at least 4,000 years ago. In the United States, fish culture had its roots as a tool in government efforts to rehabilitate depleted stocks of migrating fish, such as shad. An extensive system of federal and state freshwater sport fish hatcheries developed and became the precursor for private sport and food fish production. The practice of fish culture is part of a conglomerate of small industries now known as aquaculture.

Aquaculture is the practice of rearing, growing, or producing products in water or in managed water systems. Products of aquaculture include plants, insects, crustaceans, bivalves and pearls, fish, and anything else grown in water. Aquaculture in a marine or saltwater environment is called *mariculture*. *Pisciculture* is an early British term for fish culture. *Fish farming* is the term often used by American catfish producers.

In the United States, commercial aquaculturists note that government regulations impede the rapid development of aquaculture. On the other hand, federal and state government agencies provide support for the industry through research, extension (information and education), and the maintenance of public hatcheries that are a substantial economic base for industry suppliers. Commercial aquaculture growth has resulted in the expansion of support industries and products and in cost-effective innovations. Public culture operations benefit from private industry-generated efficiencies and technologies. Recently the academic community has experienced the growth of aquaculture-related studies—studies that did not exist before the 1970s. In addition to extension, universities conduct research and provide a platform for, and a catalyst to enhance, communication. Together, government, private, and academic views and interests are synergistic; each makes a contribution to a blossoming industry and achieves a result that is greater than the sum of its independent efforts. Each contributor gains efficiencies and derives benefits from the improvement of the whole. And a growing aquaculture industry reduces the national debt, provides employment, and enhances the quality of life.

To maximize the benefits of new technology and to operate most efficiently, good management is essential. Decision making is becoming increasingly important as intensification of systems requires larger financial investment and uses increasingly sophisticated technology. Traditionally, most culture system managers have received little management training. Current trends toward technological advances, increased capitalization, use of credit, and availability of management information and support systems, small profit margins, and more sophisticated production systems and employees all amplify the importance of decision making and planning, and are the reasons that more and better management is needed.

THE CULTURE SYSTEM SPECTRA

World Overview

Aquaculture is truly worldwide, but the greatest tradition and over 80% of the world's production are in Asia. Japan leads the world in the variety of species cultured and in the amount of marine culture production. China leads the world in the total weight of fish produced. World aquaculture production was roughly 22 billion pounds per year in 1986. The variety of species commercially produced includes at least 93 finfishes, 7 shrimps or prawns, 6 crawfish, 6 bivalves, and many (probably over 20) plants. As a portion of the world's aquatic food production, aquaculture contributed 7% in 1970 and 13% in 1980; it may account for 25% before 2010.

In the United States, absolute fish consumption increased 30% between 1970 and 1981, and 47% of the fish consumed in 1981 was imported. Per capita consumption of commercial seafood reached 15.4 pounds in 1987 and should exceed 20 pounds by 2000. The 1987 per capita commercial seafood consumption statistics did not include 3–5 pounds from aquaculture production and 3–4 pounds from recreational fishing. However the 21–24 pounds of U.S. consumption of fishery products per capita is modest compared to the 40–45 pounds in Europe and the 130–140 pounds in Japan.

Aquaculture production in the United States rapidly expanded from 205 million pounds in 1975 to over 600 million in 1986; it is forecast to exceed 2 billion pounds by 2010. The production quantity (in millions of pounds) of the top five cultured species in 1986 was: catfish (327), crawfish (98), Pacific salmon (74), trout (51), and baitfish (25). The value (in millions of dollars) of production for the top five species in 1986 was: catfish ($229), trout ($56), baitfish ($52), crawfish ($49), and oysters ($43). By 1987, Atlantic salmon production was the fastest-growing segment of U.S. aquaculture. It has been estimated that for every 10 million additional pounds of catfish produced, there are 220 jobs created within the industry and 1,100 related jobs created outside the industry. Although these statistics, primarily from Dicks and Harvey (1988), are impressive, they cannot communicate the excitement and pride generated by the people who are a part of this fast-growing industry—an industry whose rapid growth has occurred during a period of overall depression in U.S. agriculture production. Culturists in the United States have the opportunity to increase the amount and reliability of the fishery products supply, reduce the trade deficit, develop new markets for U.S. grain, provide new jobs, and increase the national revenue.

Globally, the two most widely cultured fishes are the common carp (*Cyprinus carpio*) and the tilapia (*Tilapia*) species. The common carp may be the most versatile species, grown in tropical to temperate regions in fresh and brackish water. The carp is a popular pond-reared food fish in Asia and comprises well over half of the 350 tons per year of fish reared in Europe. Carp are omnivorous and can be reared cheaply in ponds, with or without supplemental feeding, an important consideration in developing countries. The common carp has not been generally accepted as a food fish in the United States.

Tilapia are tropical to subtropical fish that can also be reared in fresh water or salt water. They are usually cultured in ponds, especially in Africa, the Middle East, and Asia, where they feed on algae and detritus. Tilapia can be reared in polyculture (with other fish or organisms) and in fish hatchery effluent waters, where they grow rapidly without supplemental feeding. However, their high reproductive rate at a small size often results in stunted populations, and monosex populations are sometimes used to eliminate reproduction.

Other carp species reared in tropical, subtropical, and temperate regions include three Chinese carps, the silver (*Hypophthalichthys molitrix*), bighead (*Aristichthys nobilis*), and grass carp (*Ctenopharyngodon idella*), and three Indian carps, the catla (*Catla catla*), Rohu (*Labeo rohita*), and mrigal (*Cirrhinus mrigala*). There are about 6 million acres of fish-rearing ponds in China and India.

Milkfish (*Chanos chanos*) and mullet (*Mugil cephalus*) are widely cultured warmwater fishes reared in brackish water. The milkfish is cultured in Africa, Asia, and the Pacific-Asian islands, and populations exist on the southwestern coast of the United States and along the coast of Mexico. The mullet, native to and cultured in Asia and India, is also cultured in Mediterranean countries.

Ornamental fishes are cultured in the United States, primarily in Florida. These fishes compose a large and varied group of over 100 species and are cultured in fresh, salt, or brackish waters. Small ponds are usually used and spawning is closely controlled, since the most valuable fish are often the colorful but recessive hybrids that make up about 25% of the fry. Ornamental fish, packed in plastic bags with water and oxygen and sealed in cardboard boxes, must tolerate rapid-freight commercial transport to retail facilities. The seven most popular families of cultured ornamental fish are the Poeciliidae (live bearers), Cyprinidae (minnows), Cyprinodontidae (killifishes), Characidae (characins), Cichlidae (cichlids), Anabantidae (gouramis), and Callichthyidae-Corydoras (catfishes).

The channel catfish (*Ictalurus punctatus*) and some other catfishes are cultured in ponds as food species, especially in the southern United States. Mississippi produces two-thirds of all the catfish in the United States. The crawfish industry is centered in Louisiana and Texas, with some activity on the East Coast. There is also crawfish production in Turkey and northern Europe. Baitfish are reared in Arkansas and across the southern United States. Striped bass, *Morone saxatilis*, and their hybrids with white bass, *M. chrysops*—the "sunshine bass"—are reared in the southeastern United States. The freshwater prawn, *Macrobrachium rosenbergii*, could be reared in the southern

United States, but recurring problems indicate that it may not be economically feasible.

Species extensively cultured in temperate climates include eels (family Anguillidae), minnows, sturgeons (family Acipenseridae), and several species of centrarchids (reared from Florida to the northern United States), such as the largemouth bass (*Micropterus salmoides*), smallmouth bass (*M. dolomieui*), pumpkinseed (*Lepomis gibbosus*), and bluegill (*L. machrochirus*). Eels are reared for food, minnows for fish bait, sturgeons for restocking of endangered species or caviar, and centrarchids for stocking recreational sport fishing ponds or lakes.

Some of the more temperate species, the coolwater fishes, are grown from the southern United States to northern Canada. These fish, reared to establish or augment existing populations in lakes, rivers, or reservoirs, include yellow perch (*Perca flavescens*), walleye (*Stizostedion vitreum vitreum*), pike-perch (*S. lucioperca*), sauger (*S. canadense*), muskellunge (*Esox masquinongy*), northern pike (*E. lucius*), and a muskellunge northern pike hybrid, the tiger muskellunge.

The salmonids are perhaps the best-known temperate freshwater cultured fish. One major group includes the Pacific salmon (*Oncorhynchus*), which die after spawning; another major group, the trouts and charrs, includes the genera *Salmo* and *Salvelinus*. Trout production accounts for a large portion of U.S. state government culture efforts. The center of commercial trout production in the United States is along the Snake River in Idaho. Many salmonids are anadromous, dividing their lives between fresh water and salt water. Sea ranching is practiced with Pacific salmon in the northwestern United States and in western Canada. The Atlantic salmon are reared in net pens along the coast in Scandinavia, especially Norway, and in Chile, Canada, and the United States.

Marine shrimp, such as *Penaeus vannamei* and *P. monodon*, are cultured in central and South America, as well as the South Pacific and Asia. Oyster, lobster, and abalone cultures are developing in the United States, and pearl oysters are cultured in Japan. Marine mussels are cultured in Europe, especially Spain, and in the United States. The seaweed Nori (*Porphyra*), cultured in Japan, accounts for one of the largest culture industries anywhere. In addition to Atlantic salmon, marine fish reared in Britain for food or stocking include turbot (*Scophthalmus maximus*), Atlantic cod (*Gadus morhua*), haddock (*Melanogrammus aeglefinus*), spotted sea trout (*Cynoscion nebulosus*), plaice (*Pleuronectes platessa*), and sole (*Solea solea*), but the intensive culture techniques for the larval phases of these species are still under development. With the exception of the Atlantic cod, which is a coldwater fish (reared at less than 10°C), the marine species are cultured in waters between 15 and 25°C. Warmwater species such as red drum (*Sciaenops ocellatus*) can be cultured in the southern United States.

Fresh Water

Various sources of fresh water are used for culture (Table 1-1). The temperatures of surface waters—streams, rivers, and lakes—fluctuate with ambient air temperatures. Surface waters may vary in quality with the weather or season; they are exposed to pollutants and contaminants, and they usually contain fish and other aquatic organisms that can be reservoirs of disease. Groundwater sources—springs or wells—are most commonly used for culture. Groundwater from wells requires pumping. Free-flowing springs and artesian supplies eliminate pumping costs, the risk of pump failure, and the power backup requirement.

A profile of the source water quality and flow (over a year or a production cycle period) should be completed prior to the construction of a rearing facility. Although healthy wild populations of fish may indicate that the water supply is suitable for rearing, endemic fish may have adapted to a detrimental condition and may not complete all phases of the life cycle in that location. Quality standards for culture water are listed in Table 1-2.

Table 1-1. Characteristics of freshwater sources for fish culture

WATER SOURCE	ADVANTAGES	DISADVANTAGES
Lakes and reservoirs	1. Large volume available for special or seasonal needs 2. Intakes at two levels give temperature control	1. Susceptible to climatic changes and pollution 2. Pathogens may be present
Streams or shallow springs	1. Temperatures are usually optimum for native fish 2. Usually have high oxygen content	1. Highly variable chemical quality and sediment load due to climatic influences 2. Susceptible to pollution 3. Pathogens may be present
Deep springs	1. Nearly constant flow, quality, and temperature 2. Usually sediment free 3. Little effect of drought	1. Oxygen may be low 2. Supersaturation of nitrogen
Wells	1. Small area needed for development 2. Advantages similar to those of deep springs	1. Yield difficult to predict before development 2. Pumping costs; power backup required 3. May deplete groundwater resources 4. Supersaturation of nitrogen

Source: Adapted from Buss 1979.

Table 1-2. Water quality standards for fish culture

Alkalinity (as $CaCO_3$)	10–400
Aluminum (Al)	<0.01
Ammonia (NH_3)	<0.02
Arsenic (As)	<0.05
Barium (Ba)	5
Cadmium	
Alkalinity<100 ppm	0.0005
Alkalinity>100 ppm	0.005
Calcium (Ca)	4–160
Carbon dioxide (CO_2)	0–10
Chlorine (Cl)	<0.003
Chromium (Cr)	0.03
Copper	
Alkalinity<100 ppm	0.006
Alkalinity>100 ppm	0.03
Dissolved oxygen (DO)	5 mg/L to saturation
Hardness, Total	10–400
Hydrogen cyanide (HCN)	<0.005
Hydrogen sulfide (H_2S)	<0.003
Iron (Fe)	<0.1
Lead (Pb)	<0.02
Magnesium (Mg)	<15
Manganese (Mn)	<0.01
Mercury (Hg)	<0.2
Nitrogen (N)	<110% total gas pressure, <103% as nitrogen gas
Nitrate (NO_3)	0–3.0
Nitrite (NO_2)	0.1 in soft water
Nickel (Ni)	<0.1
PCB (polychlorinated biphenyls)	0.002
pH	6.5–8.0
Potassium (K)	<5.0
Salinity	<5%
Selenium (Se)	<0.01
Silver (Ag)	<0.003
Sodium (Na)	75
Sulfate (SO_4)	<50
Sulfur (S)	<1.0
Total dissolved solids (TDS)	<400
Total suspended solids (TSS)	<80
Uranium (U)	<0.1
Vanadium (V)	<0.1
Zinc (Zn)	<0.005
Zirconium (Z)	<0.01

Source: Wedemeyer 1977; U.S. Environmental Protection Agency 1979–80; Piper et al. 1982.

Note: Values are in milligrams per liter unless otherwise noted.

Salt Water

Important site restrictions apply to marine and estuarine systems concerning weather, pollution, and government regulations. Two types of finfish culture systems are often used—net-pen culture, to grow fingerling fish to market size, and sea ranching, in which salmon smolts are released at a size of 14–20 cm in length and captured as mature fish when they return to spawn.

Water of intermediate salinity is used for the production of estuarine fish and oysters and in soft-shell crab "shedding" procedures. Mullet and milkfish are well-known brackish-water culture species. Brackish water culture techniques have been developed for carps, tilapia, salmonids, and striped bass.

Temperature

Rates and efficiencies of feeding, digestion, and growth depend upon temperature. Each species has a characteristic growth curve that changes with temperature and size, and each has a temperature range, bounded by an upper and a lower lethal limit, beyond which it cannot survive. There is also an optimum temperature range for growth (Fig. 1-1). Within a species' tolerable temperature range, the growth rate will gradually reach a maximum level, and then decline just before the upper lethal limit is reached. The optimum temperature for growth is close to the preferred temperature or that selected by the organism if given the choice. However, the temperature preferendum curve changes with the developmental stage over the life cycle. A fish's immune response, or its ability to ward off disease, is best near the optimum growth temperature. The probability of culture success is also enhanced at or near the temperature for optimum growth. As temperatures rise above the optimum, tolerance to metabo-

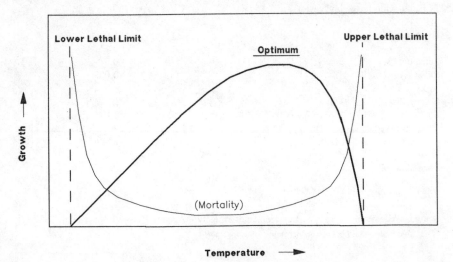

Figure 1-1. Relation of fish growth to temperature.

lites decreases and intensification of culture must be reduced; for instance, more water flow is required and feeding must be reduced to reduce metabolites and maintain fish health. Some of the reported optimum growth temperatures for fish are given in Table 1-3.

Organizational Systems

Subsistence culture is the rearing of fish for personal consumption; it might be called fish "gardening." There are two general types—low-technology, extensive pond systems and small-container, intensive systems. Each is designed to produce a few hundred pounds of edible fish in a growing season. Pond culture methods usually involve little cost other than that for pond construction, and typically an existing recreational pond is used. Cropping is often accomplished by angling but can also be done by seining or draining at the end of a growing season, or cages can be used to confine the fish during culture. More intensive systems may involve a tank, such as a child's swimming pool, and a recirculating pump. The species reared must be tolerant to the temperatures experienced over the growing period, but some warmwater species can be reared in northern climates during summer and some coldwater fishes can be reared in southerly climates over winter. There are purchase, maintenance, and operating costs involved, but many fishes can be reared in little space.

Until recently, much of the commercial fish production in the United States and Canada was done by small family-owned businesses. Small businesses continue to rear fish for local sale as food or bait, for stocking private waters, or for fee fishing. Small business operations are sometimes part of an organized group or cooperative, in which several growers work together to gain some purchasing or marketing advantage.

Some large culture operations are subsidiaries of much larger, diversified corporations. These operations tend to have a high capital development in the culture facilities. Large businesses often have some vertical integration of enterprises, such as fish food production or fish processing operations, and may have research and distribution components. Large U.S. businesses produce trout, salmon, channel catfish, and crawfish within the United States and operate extensive shrimp farms in other countries.

Government culture operations occur at the local, state or provincial, and federal government levels. Unlike the commercial or subsistence operations, most government production is not for table fare, but is done to provide fish for stocking public recreational waters or to reestablish depleted or endangered wild stocks.

Operational Systems and Degree of Intensification

Culture systems are usually labeled extensive or intensive and use either static or flowing water. Extensive, static pond culture is the oldest form of fish culture. In more intensive pond production, water is flushed through the system (as in

Table 1-3. Estimated optimum rearing temperatures of cultured species

COMMON NAME	SCIENTIFIC NAME	OPTIMUM TEMPERATURE RANGES, °C
Freshwater		
American shad	*Alosa sapidissima*	7–23
Pink salmon	*Oncorhynchus gorbuscha*	9–17
Lake trout	*Salvelinus namaycush*	10–15
Sockeye salmon	*Oncorhynchus nerka*	10–17
White bass	*Morone chrysops*	10–18
Brown trout	*Salmo trutta*	10–18
Coho salmon	*Oncorhynchus kisutch*	11–17
Chinook salmon	*Oncorhynchus tshawytscha*	12–17
Rainbow trout	*Salmo gairdneri*	13–21
Atlantic salmon	*Salmo salar*	14–18
Brook trout	*Salvelinus fontinalis*	15–20
Golden shiner	*Notemigonus crysoleucas*	17–24
Smallmouth bass	*Micropterus dolomieui*	18–24
Sauger	*Stizostedion canadense*	19–22
Northern pike	*Esox lucius*	19–26
Yellow perch	*Perca flavescens*	20–27
Walleye	*Stizostedion vitreum vitreum*	20–23
Eels	*Anguillidae*	20–28
Striped bass	*Morone saxatilis*	22
Fathead minnow	*Pimephales promelas*	23–29
Brown bullhead	*Ictalurus nebulosus*	23–31
Muskellunge	*Esox masquinongy*	24
Goldfish	*Carassius auratus*	24–30
Channel catfish	*Ictalurus punctatus*	25–30
Pumpkinseed	*Lepomis gibbosus*	26–31
Guppy	*Poecilia reticulata*	27–29
Mosquitofish	*Gambusia affinis*	27–31
Largemouth bass	*Micropterus salmoides*	27–32
Tilapia	*T. nilotica or T. mossambica*	28–30
Common carp	*Cyprinus carpio*	28–32
Bluegill	*Lepomis machrochirus*	29–32
Saltwater		
Atlantic cod	*Gadus morhua*	1–9
Plaice	*Pleuronectes platessa*	16–17
Sole	*Solea solea*	20–23
Roach	*Rutilus rutilus*	27

Source: Bardach et al. 1972; Coutant 1977; McCauley and Casselman 1981; and Jobling 1981.

crawfish production) or aeration is routinely used (as in catfish production). In the system often referred to as *flowing water intensive culture*, fish are reared at relatively high densities in a container supplied with flowing water.

Intensification is the process of increasing the metabolic load, or the biomass if all else is equal, on a unit of water flow. High-intensity systems may use devices for oxygenation and removal of metabolites. The array of systems has, on the extensive extreme, a simple pond with no aeration and little or no outflow, and, on the intensive extreme, rearing containers with large volume inflows effecting frequent, complete water changes. Supplemental oxygenation, and perhaps water reuse after reconditioning of the water through a biological filter and a sterilization procedure, may also be used in intensive systems. A gradient of intensification is described in Figure 1-2, and relates the increase in density with increasing inputs of feed, aeration, and water flow. The risk of loss increases proportionally with intensification, due to the inherent dependence on life support technology and the increased potential for contact with disease organisms. Losses due to oxygen deficiencies can occur in pond systems, traditionally referred to as extensive culture (Fig. 1-2), when fish production reaches the range of 2,000–5,000 kg/ha. Risks of intensification can be reduced by backup of the life support systems, which should be planned as an integral part of the intensive culture operation.

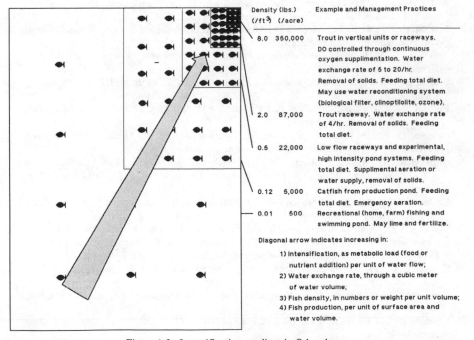

Figure 1-2. Intensification gradient in fish culture.

SUMMARY

The world's fishing industry has approached its limits on the harvest of traditional food fishes. In a growing number of situations, culture is a more effective means of producing an aquatic product. More capitalization, more technology, more use of credit, more sophisticated production systems, and decreased profit margins force more effective management.

The U.S. aquaculture industry is relatively small and new compared to that of China and Japan. Freshwater, estuarine, and marine culture environments are distinctly different. Water temperature also limits the choice of species and strongly affects their growth and health. In some parts of the world, aquaculture is practiced for subsistence. In the United States, the culture industry is composed of both small and large businesses. Some large U.S. businesses have developed their aquaculture operations outside the United States. Intensification refers to the biological load on a unit flow or volume of water. The risk is related to the degree of intensification. High intensification usually involves high water flow or oxygen supplementation.

REFERENCES AND RECOMMENDED READINGS

Bardach, J. E., J. H. Rhyther, and W. O. McLarney. 1972. Aquaculture. New York: Wiley-Interscience.

Bell, F. W. and E. R. Canterberry. 1976. Aquaculture for the Developing Countries: A Feasibility Study. Cambridge, MA: Ballinger Publishing Company.

Bell, M. C. 1972. Fisheries Handbook of Engineering Requirements and Biological Data. Portland, OR: U.S. Army Corps of Engineers.

Brown, E. E. 1977. World Fish Farming: Cultivation and Economics. Westport, CT: AVI Publishing Company, Inc.

Brown, E. E. and J. B. Gratzek. 1980. Fish Farming Handbook. Westport, CT: AVI Publishing Company, Inc.

Buss, K. 1979. The Fundamentals of Fish Culture: An Outline for Classroom Study. Mansfield, PA: Mansfield State College Press.

Caturano, S., L. S. Glanz, D. C. Smith, L. Tsomides, and J. R. Moring. 1988. Shellfish mariculture: The status of mussel power in Maine. Fisheries 13(3):18–21.

Chamberlain, G. W., R. J. Miget, and M. G. Habey. Editors. 1987. Manual on Red Drum Aquaculture. College Station, TX: Texas Agricultural Extension Service and Sea Grant College Program.

Coche, A. G. 1985. A list of selected FAO publications related to aquaculture, 1966–1985. FAO Fisheries Circular 744, Revision 1.

Colt, J. 1987. An introduction to water quality management in intensive aquaculture. In Oxygen Supplementation: A New Technology in Fish Culture, ed. L. Visscher and W. Godby. U.S. Dept. of the Interior, Fish and Wildlife Service, Region 6, Information Bulletin #1.

Coutant, C. C. 1977. Compilation of temperature preference data. J. Fish. Res. Board Can. 34:739–745.

Dicks, M. and D. Harvey. 1988. New industry fishes for acceptance. Agricultural Outlook. June/AO142; 16–17.

Edwards, D. E. 1978. Salmon and Trout Farming in Norway. Farnham, Surrey, England: Fishing News Books Ltd.

Godfriaux, B. L., A. F. Eble, A. Farmanfarmaian, C. R. Guerra, and C. A. Stephens. 1979. Power Plant Waste Heat Utilization in Aquaculture. Montclair, NJ: Allanheld, Osmun and Company.

Gordon, M. R., K. C. Klotins, V. M. Campbell, and M. Cooper. 1987. Farmed Salmon Broodstock Management. Vancouver, Canada: B.C. Research.

Hepher, B. and Y. Pruginin. 1981. Commercial Fish Farming. New York: John Wiley & Sons, Inc.

Huner, J. V. and E. E. Brown. 1985. Crustacean and Mollusk Aquaculture in the United States. Westport, CT: AVI Publishing Company, Inc.

Iversen, E. S. 1968. Farming the Edge of the Sea. London: Fishing News Books Ltd.

Jobling, M. 1981. Temperature tolerance and final preferendum—rapid methods for the assessment of optimum growth temperatures. J. Fish Biol. 19:439–455.

Krise, W. F. and J. W. Meade. 1986. Review of the intensive culture of walleye fry. Prog. Fish-Cult. 48(2):81–89.

Leitritz, E. and R. C. Lewis. 1976. Trout and Salmon Culture. Fish Bulletin 164. Long Beach, CA: State of California Department of Fish and Game.

Limburg, P. R. 1980. Farming the Waters. New York: Beaufort Books, Inc.

Logsdon, G. 1978. Getting Food from Water: A Guide to Backyard Aquaculture, Emmaus, PA: Rodale Press.

McCauley, R. W. and J. M. Casselman. 1981. The final preferendum as an index of the temperature for optimum growth in fish. In Proceedings of the World Symposium on Aquaculture in Heated Effluents and Recirculation Systems, ed. Klaus Tiews, pp. 82–93. Berlin: H. Heenemann.

McLarney, W. and J. Parkin. 1981. Back Yard Fish Farm Book. Andover, MA: Brick House Publishing Company.

McLarney, W. 1984. The Freshwater Aquaculture Book. Point Roberts, WA: Hartley and Marks.

Meade, J. W. and C. A. Lemm. 1986. Effects of temperature, diet composition, feeding rate, and cumulative loading level on production of tiger muskellunge. In Managing Muskies: A Treatise on the Biology and Propagation of Muskellunge in North America. Am. Fish. Soc. Special Publ. 15:292–299.

Meade, J. W., W. F. Krise, and T. Ort. 1983. Effect of temperature on production of tiger muskellunge in intensive culture. Aquaculture 32(1–2):157–164.

Meade, J. W., J. S. Ramsey, and J. C. Williams. 1985. Effects of cumulative loading level, as fish weight per unit flow, on water quality and growth of lake trout. J. World Maricul. Soc. 16:40–51.

Michael, G. R. 1987. Managed Aquatic Ecosystems of the World. New York: Elsevier Science Publishing Company.

Milne, P. H. 1972. Fish and Shellfish Farming in Coastal Waters. London: Fishing News Books Ltd.

Perry, W. G. and J. Tarver. 1987. Polyculture of Macrobrachium rosenbergii and Notemigonous crysoleucas. J. World Aquac. Soc. 18(1):1–5.

Pillay, T. V. R. and W. A. Dill. 1979. Advances in Aquaculture. Farnham, Surrey, England: Fishing News Books Ltd.

Piper, R. G., I. B. McElwain, L. E. Orme, J. P. McCraren, L. G. Fowler, and J. R. Leonard. 1982. Fish Hatchery Management. Washington, DC: U.S. Fish and Wildlife Service.

Pullin, R. S. V. and Z. H. Shedadeh, eds. 1980. Integrated Agriculture-Aquaculture Farming System. ICLARM Conference Proceedings 4. Los Banos, Laguna Philippines International Center for Living Aquatic Resources Management, Manila and the Southeast Asian Center for Graduate Study and Research in Agriculture.

Rhodes, R. J. 1986. Aquaculture developments in the Americas. Infofish Marketing Dig. 4:9–11.

Rhodes, R. J. 1988. Status of world aquaculture. Aquaculture Magazine, 18th Annual Buyer's Guide, pp. 6–20.

Sedgwick, S. D. 1985. Trout Farming Handbook. Farnham, Surrey, England: Fishing News Books Ltd.

Senn, H., J. Mack, and L. Rothfus. 1984. Compendium of Low-Cost Pacific Salmon and Steelhead Trout Production Facilities and Practices in the Pacific Northwest. Portland, OR: U.S. Dept. of Energy.

Sindelar, S. 1984. Potentials in aquaculture. In Biotechnology in the Marine Sciences, Proceedings of the First Annual MIT Sea Grant Lecture and Seminar, ed. R. R. Colwell, A. J. Sinskey, and E. R. Pariser. New York: John Wiley & Sons, Inc.

Smith, L. J. and S. Peterson, eds. 1982. Aquaculture Development in Less Developed Countries: Social, Economic, and Political Problems. Boulder, CO: Westview Press.

Speece, R. E. 1973. Trout metabolism characteristics and the rational design of nitrification facilities for water reuse in hatcheries. Trans. Am. Fisheries Soc. 102:323–334.

Stauffer, J. R., Jr., E. L. Melisky, and C. H. Hocutt. 1984. Interrelationships among preferred, avoided, and lethal temperatures of three fish species. Arch. Hydrobiol. 100:159–169.

Stickney, R. R. 1986. Culture of Nonsalmonid Freshwater Fishes. Boca Raton, FL: CRC Press, Inc.

U.S. Department of Agriculture. 1987. Aquaculture: A Guide to Federal Government Programs. Prepared by the Joint Subcommittee on Aquaculture, in cooperation with the National Agricultural Library, USDA. Washington, DC: USDA.

U.S. Environmental Protection Agency. 1979–80. Water Quality Standards Criteria Digest: A Compilation of State/Federal Criteria. Washington, DC: U.S. EPA.

Watson, A. S. 1979. Aquaculture and Algae Culture. Park Ridge, NJ: Noyes Data Corporation.

Webb, P. V. 1988. Partitioning of energy into metabolism and growth. In Ecology of Freshwater Fish Production, ed. S. D. Gerking, pp. 184–214. New York: John Wiley & Sons, Inc.

Webber, H. H. 1984. Aquabusiness. In Biotechnology in the Marine Sciences, Proceedings of the First Annual MIT Sea Grant Lecture and Seminar, ed. R. R. Colwell, A. J. Sinskey, and E. R. Pariser, pp. 115–122. New York: John Wiley & Sons, Inc.

Wedemeyer, G. A. 1977. Environmental requirements for fish health. In Proceedings of the International Symposium of Diseases of Cultured Salmonids, pp. 41–55. Seattle: Travolac, Inc.

Wellborn, T. L., Jr. 1987. Catfish Farmer's Handbook. Mississippi Cooperative Ext. Serv. Publ. 1549.

Westers, H. 1979. Controlled fry and fingerling production in hatcheries. EIFAC Workshop on Mass Rearing of Fry and Fingerlings of Fresh Water Fishes, ed. E. A. Huisman and H. Hogendorn, pp. 32–52. EIFAC/T35, Supplement 1. Hague: Ministry of Agriculture and Fisheries.

Westers, H. 1984. Principles of Intensive Culture. Lansing, MI: Michigan Dept. of Natural Resources.

Westers, H. and K. M. Pratt. 1977. Rational design of hatcheries for intensive salmonid culture, based on metabolic characteristics. Prog. Fish-Cult. 39:157–165.

Chapter 2
Principles of Culture Systems Management

MANAGEMENT DEFINITIONS AND APPROACHES

Management is getting what you want by using what you have. Formally, management is the planned, wise use of available human, capital, equipment, material, and environmental resources to accomplish goals. For purely commercial purposes it might be defined as the organization and operation of an aquatic culture business from the viewpoint of efficiency and continued profit.

Approaches to management usually relate to the job, to resources, or to problem solving. From one management situation to another, the job of running the operation does not change in concept, yet it usually allows great latitude in style. Resources are interchangeable, at least to some extent, through management. The problem-solving, or decision-making, approach is dynamic, since each problem is a new or an unanswered question, and each is an opportunity.

THE JOB OF MANAGEMENT

Management in an organization is somewhat analogous to the nerve center in an organism. Whereas an owner or board of directors may establish goals, the manager sets objectives, makes or facilitates major decisions, and accepts responsibility for accomplishment of the objectives. A manager uses available resources, including labor (people), capital, equipment, and materials to reach the objectives that meet goals. Typically managers continuously perform a great number of brief and varied activities. They keep things rolling, get more things started, identify and correct problems, and generally "watch the store." However, they may do these things indirectly by managing managers, that is, each subordinate may be considered a manager of an area of responsibility. Important aspects of the manager's job have been studied and described in detail.

MANAGEMENT FUNCTIONS AND ACTIVITIES

The five management functions—planning, organizing, staffing, directing, and controlling—and the activities associated with each are listed in Table 2-1. Each function and activity is performed by every professional manager at every level of management. Differences exist in the magnitude and frequency with which they are performed. The greatest change in effort is in the directing function, as the closer the manager is to production activity, the more effort is devoted to directing. As a manager proceeds up the ladder, a marked change in the management functions and activities mix becomes inescapable. Thus a highly successful first-line supervisor may not necessarily make a good middle manager, and vice versa.

Table 2-1. Functions and activities of management work as synthesized by Morrisey (1970)

Function I. Planning. Determining the work that must be done.

1. Defining roles and missions, and the nature and scope of work.
2. Forecasting. Estimating the future.
3. Setting objectives. Determining results to be achieved.
4. Programming. Establishing a plan of action to meet objectives.
5. Scheduling. Establishing time requirements for objectives.
6. Budgeting. Determining and assigning the resources required to reach objectives.
7. Policy making. Establishing rules, regulations, or predetermined decisions.
8. Establishing procedures. Determining consistent and systematic methods of handling work.

Function II. Organizing. Classifying and dividing the work into manageable units.

9. Structuring. Grouping work for effective and efficient production.
10. Integrating. Establishing conditions for effective teamwork among organizational units.

Function III. Staffing. Determining the requirements for, and ensuring the availability of, personnel to perform the work.

11. Determining personnel needs. Analyzing the work for personnel capabilities required.
12. Selecting personnel. Identifying and appointing people to organizational positions.
13. Developing personnel. Providing opportunities for people to increase their capabilities in line with organizational needs.

Function IV. Directing (leading). Bringing about the human activity required to accomplish objectives.

14. Assigning. Charging individual employees with job responsibilities or specific tasks.
15. Motivating. Influencing people to perform in a desired manner.
16. Communicating. Achieving an effective flow of ideas and information in all desired directions.
17. Coordinating. Achieving harmony of the group effort toward the accomplishment of individual and group objectives.

Function V. Controlling. Ensuring the accomplishment of objectives.

18. Establishing standards. Devising a gauge of successful performance in achieving objectives.
19. Measuring performance. Assessing actual versus planned performance.
20. Taking corrective action. Bringing about performance improvement toward objectives.

Planning and Organizing

Fulmer (1974) defined planning as the primary function of a manager and described steps to be included in a thorough planning process.

1. Choose objectives—clear-cut, carefully considered statements of the end points of planning.
2. Communicate objectives.
3. Identify premises or assumptions and assess the future. (The Delphi process, discussed in Chapter 11, can be a useful tool.)
4. Survey resources, reconsidering proposals in light of the actual situation.
5. Establish policies. Policies are shortcuts for thinking that help to solve problems in advance and provide a foundation for subsequent decisions.
6. Identify alternatives. Without alternatives, planning is unnecessary.
7. Compare alternatives.
8. Choose a course of action. Avoid the decision not to decide.
9. Create procedures and rules. A procedure is a step-by-step action guide within the policy framework. Rules have no sequence and typically stand alone.
10. Establish budgets. Include all resources that will be consumed.
11. Establish timetables. Include milestones as well as completion dates.
12. Establish standards. Control is impossible without planning.

A special situation in planning and organizing involves staffing and position management. In theory, the manager first establishes operational goals and a strategy to achieve those goals. The strategy dictates an organizational structure, which is then staffed. The manager delegates authority and responsibility and monitors progress, making changes when, and only when, necessary. Reorganization can be helpful, but it is nearly always traumatic and involves a risk of time and productivity. Thus it is important to invest sufficient resources in the early planning and staffing stages. Once the organization is staffed, it develops a personality of its own, and from then on the organization is built around people.

It is difficult to be most effective when second-rate staffing and planning are used. Although the manager must get the most out of the staff, the manager must first get the most out of the staffing plan by building a plan that enhances efficiency and productivity. The staffing plan is an important first step in the dynamic area of position management. Here are seven principles of position management:

1. Delineate clearly the work assignment of each position.
2. Delineate clearly the job-to-job relationships.

3. Allocate responsibilities so that higher-level duties are not spread thinly among positions.
4. Include sufficient challenge, variety, and responsibility in positions to attract and retain able employees.
5. Include well-defined career ladders, providing candidates for key positions in the organization.
6. Balance the proportions of senior, junior, and support positions.
7. Ensure a small proportion of managerial and supervisory positions to nonsupervisory positions, and avoid layering of supervisory positions.

Decision Making

Some managers believe that decision making is the essence of management, or perhaps that decision making is management. Decision making often involves planning, just as planning involves decision making. Osburn and Schneeberger (1983) suggested that the decision-making steps should follow the scientific method for problem solving (modified below).

1. Clearly define the problem, or set the goals or objectives.
2. Gather and organize objective facts and data.
3. Analyze and evaluate the facts and data.
4. Draw conclusions or make the decision.
5. Take action.
6. Accept responsibility for the decision.
7. Evaluate the outcome.

Control and Evaluation

Some of the most successful control techniques are subtle and nearly unnoticeable. The three parts of control are standard, measurement, and redirection. Setting the standard involves determining and communicating what needs to be done. Development of worthwhile standards requires and deserves substantial effort by the manager. All standards should be easily measurable and based on results rather than methods. Redirection is the development and communication of the next standard, which may involve correction for earlier performance.

Measurement is the evaluation of performance, that is, determining if that which needed to be accomplished was, in fact, accomplished. It might be said that anything worth doing is worth evaluating. Evaluation is the feedback or information circuit in the control mechanism; it keeps the process on track and is essential for, and part of, control. Evaluation is not a necessary evil, but a helpful and healthy step used to increase efficiency and productivity.

Evaluation of worker performance is usually done by comparing the actual performance to the stated standard. The evaluation of processes also uses standards; however, understanding the effectiveness of approaches may require more sophisticated analyses. Operations research is a field of study that offers methods to enhance the planning and redirection of serial processes. Computers have made the power of linear programming readily available and can be applied to nearly any production system.

Accepting Responsibility

Acceptance of responsibility is an immediate distinction between worker and manager. If the manager is not responsible, the manager is either not needed or is actually only an advisor. Risk is involved in every decision. If outcomes were always known, there would be no need for decision makers. Therefore, undesirable outcomes sometimes follow "good" decisions. The manager, not a hired fish culturist, is responsible for an undesirable outcome.

MANAGEMENT STYLES AND ATTRIBUTES OF SUCCESSFUL MANAGERS

If the mean annual discounted cost of employing a manager over a 20-year period is $50,000, the organization invests $1 million in the manager. The organization should invest wisely and strive to obtain the best management available for that cost. Likewise the manager has an obligation to the organization to provide a continuing high level of effectiveness. Three generalized management models are described in Table 2-2. Management style is closely associated with the experience, personality, abilities, and aspirations of the individual manager. Familiarity with an array of styles, coupled with an outline of the attributes common to many successful managers, may help the developing manager to become more effective. What are important aspects of manager or executive selection? Are there attributes common to all successful managers? How should a manager or aspiring manager maintain, update, and hone the skills required to remain in the forefront of effective management?

The most important criteria for selection of professionals, reported in a survey of 252 personnel directors of Fortune 500 companies (Associated Press release, Washington, D.C., January 12, 1988), were experience, competence, and the absence (versus presence) of any hint of marijuana or drug use. Although experience is almost always deemed important, these criteria contrast with the traditional view that selection criteria are often based on the college attended, personal appearance, fitness and weight, tobacco use, and personality. Although academic credentials, personal appearance, and personal habits have aesthetic appeal and may be very important for public image and relations, they appear to be window dressing compared with the bottom-line management aspects of experience and competence. Only 2% of the personnel directors said that they were "very unlikely" to hire an otherwise qualified candidate who drinks

Table 2-2. Management models

TRADITIONAL MODEL	HUMAN RELATIONS MODEL	HUMAN RESOURCE MODEL
Close supervision of subordinates performing narrowly defined jobs. People are lazy, uncreative, and concerned only with what they earn, not with what they do to earn it. Adequate performance is achieved only through tight control.	Limited participation of subordinates in decision making and self-control. People want to be useful or important, and have high belonging or social needs that are important motivators to work and cooperate. Limited participation improves morale.	Ever-expanding participation and self-direction of subordinates. Employees' abilities not fully used; behave responsibly in achieving goals they help set. Participation improves effectiveness and satisfies deeper needs.
Management functions: plan, organize, staff, direct, control.	Management functions: attitudes control production. Potency of informal group. Worker has feelings and is not just a creature with economic needs.	Management functions: Encourage full participation and collaboration. People like to work, are responsible and creative, and have high achievement needs.
Management principles: division of work, unity of command, chain of command, line and staff. Theory X. Traditional view of direct and control/job simplification.	Management Principles: Theory X: Happy workers are productive workers.	Management Principles: Theory Y: Integration of organization and individual needs. Job enrichment.
Stick approach. Satisfies subsistence needs. Organizational/task concerns.	Carrot approach. Satisfies social and belonging needs. Relationship concerns.	Satisfies esteem and self-actualization needs.

after work, whereas 47% were "very unlikely" to hire an applicant who uses marijuana after work. The disparity between disqualification due to marijuana use versus alcohol use may be an age-related social bias, but it is a current reality.

Identification of the characteristics and attributes commonly shared by successful people can provide insight and direction for those interested in personal or staff improvement. Successful individuals look for new challenges rather than rest on a plateau, initiate unassigned projects rather than mark time, try immediately to correct a problem rather than assign blame, compete inwardly rather than with others, and become prepared for an event by "brainstorming" alone beforehand so as to encounter the event and proceed confidently. Attributes that were most frequently found to be characteristic of successful leaders were listed by Yukl (1981). Additional attributes are listed in Appendix I.

The selection of supervisors and managers in the U.S. Fish and Wildlife Service is based on (1) technical competence, (2) loyalty to the Fish and Wildlife

Service, (3) ability to work with people, and (4) knowledge of the organization. However, many believe that the single most important reason for any promotion involves being in the right place at the right time. A listing of the expectations of an entry-level manager in the U.S. Fish and Wildlife Service can be found in Appendix I.

Likert (1967) conducted a 10-year study of "Productivity, Supervision, and Employee Morale," with the cooperation of several large companies. Part of the findings dealt with the way employees described supervisors who had been identified (by their supervisors) as "immediately promotable." A summary of the way employees characterized such supervisors can be found in Appendix I. Likert wrote: "The notion of a 'good' supervisor as the strong man who makes decisions independently is not borne out. The data point towards the importance of group functioning." Likert also indicated that low-production supervisors felt that they did not have the information or the authority to act competently. "The more supervisors participate in decisions affecting them and their work, the greater is the extent to which they use participation and group decision with their own work group. These results are in sharp contrast with the not uncommon idea that each level of organization should keep a sharp eye and close control over the action of those at the next lower level."

SCOPE OF AQUACULTURE MANAGEMENT

Although the management of one or more enterprises within an aquaculture organization may be accomplished by a single person, or by a group or team of specialists, the scope and nature of the management functions are similar. Areas that must be managed can be categorized as technical, personnel, financial, and accounting activities. In turn, these activities may require the manager to be a goal setter, planner, designer, purchaser, supervisor, motivator (counselor), advertiser, marketing agent, public relations person, and more. Fisheries has been called a trans-disciplinary science, and aquaculture involves not only fisheries but also general biology, nutrition, chemistry, statistics, and economics. And all these parts come into play in management. Table 2-3 gives a sampling of the scope of management activities.

BOTTOM LINE

Managing to Achieve Objectives

Bottom-line management implies focusing on the objective. Many new management approaches, with high-impact names, are variations of a system developed for industry in the late 1950s known as *management by objectives* (MBO). MBO is an approach to planning and achieving long-range goals; it is also used for control. In fact, the concept is useful for most management activities. A goal is reduced to several objectives or components; then each objective is addressed through the planning process. The emphasis in MBO is on the objective as

Table 2-3. A sampling of management activities

Technical

Water use	Production alternatives
	Pre- and postuse treatment
	Constraints and regulations
	Water quality determination and maintenance
What and how to produce	Enterprise selection and mix
	Production schedules and scale
	Resource or input requirements
	Quality
Level of mechanization	Intensification tradeoffs
	Capital required
	Services available
	Anticipated maintenance
Production problems and diseases	Potentials
	Diagnostic capabilities
	Diagnostic services available
	Correction (and treatment) capabilities

Personnel

Staffing	Alternatives
	Recruiting and hiring
	Reductions
	Payroll; time and service remuneration
Training	Pursue or avoid
	On-the-job
	Specialties
	Backup
Productivity	Management styles
	Supervision
	Motivation
People opportunities	Team mix
	Rewarding efforts
	Discipline and removal
	Counseling

Financial

Funding	Acquiring funds
	Using funds
	Forecasting needs
Purchasing and contracting	System
	Budgeting
	Qualities and quantities
	Monitoring
Marketing	How; mix of marketing and advertising approaches or efforts
	Who, how much
	Value added
	Research

Table 2-3 (*cont.*)

Accounting	
Production records	Enterprise
	Resources (water, cost, labor)
	Production (rearing) unit
	Others
Business transactions	Accounting method
	Accounting summaries and reports
	Cash flow and forecasting
Tax reporting	Income (and similar taxes)
	FICA
	Depreciation
	Other
	Filing documents

the bottom line, and the success of the effort, or the evaluation of the person responsible, is based on the degree to which the stated objective was met. An objective must be:

Doable
Affordable
The responsibility of one person (somebody must be in charge)
Adaptable to the timetable
Negotiable
Nonconflicting
Measurable

Details on developing MBO for employee performance are given in the discussion of control and evaluation in Chapter 3.

MBO is not a panacea. It is a proven system that depends on, but also promotes, vertical communication. MBO requires an understanding, developed over time, among those at several organizational levels. As with many systems, simplicity enhances its effectiveness. Making the employee an integral part of the planning and evaluation process opens the door to such motivators as commitment, job satisfaction, teamwork, self-esteem, and autonomy. The employee "buys into" the objective and "owns" the responsibility. The employee has the incentives of changing the outcome and earning a reward for a known level of performance.

Financial Analyses Miss the Point in the Public Domain

The human resource application of MBO is straightforward and sensible. MBO is also used for process and operations management. However, a special situation arises with public sector managers regarding financial accountability.

Should such managers address their objectives from a cost point of view, as "for profit" managers might? Many public sector managers say they should not, since that approach diverts attention from the purpose of their efforts and, in fact, misses the point entirely. For instance, the public sector manager may use several approaches, if mandated, to restore Atlantic salmon to a New England river in which the nursery and spawning areas are effectively cut off from ocean fish migration due to the presence of a hydroelectric dam. One approach would be to rear salmon on, or for release in, the river to establish a returning or sea-run population. Another would be to facilitate fish passage at the dam. A fish passageway may cost $10 million. A fish hatchery may cost $5 million and require $0.5 million per year to operate. Substantial results cannot be expected from the hatchery for 10 years.

If after 5 years of operation there is an annual average of 10 hatchery-reared salmon returning from the sea to the river, the dam owner may suggest that it costs $50,000 per fish in operating costs alone to produce those returning salmon. Proponents of the dam may further assert that since salmon can be purchased for, say, $10 per pound, this hatchery production and fish ladder construction is a ridiculous approach, and the cost of a fish ladder cannot be justified to pass $500 worth of fish. The suggestion might be made to simply purchase fish elsewhere and forget the hatchery and fish ladder.

The public sector manager could not counter the criticism by defending the costs as attributable to fish production. That is, if the object of the hatchery production was to produce a certain number of pounds of fish, the cost argument would be valid. However the manager's object is not those 10 returning fish, any more than it was the zero fish that returned during the first 2 years of hatchery operation or the 500 fish that might return in a following year. If costs are to be considered, they must address the entire picture and the point of the restoration—that is, the lost resource and all of its benefits.

The resource that was lost accounted for extensive recreational fishing activities. Those activities involved not just fishing hours but all associated expenditures for such things as equipment, transportation, food and lodging, and other spending by fishermen. By today's standards, such a resource generates significant economic activity. There also may have been commercial fishing aspects associated with the resource; any loss of salmon attributable to construction of dams indicates that other migrating fish were lost to the river system. And there is inherent value, placed by the public and by Congress, on the population or species being restored. ("Willingness to pay" evaluations often reveal a surprisingly high value placed by the public on a natural resource.) Surely the dam operator is not suggesting, through his criticism, that he might be responsible for all that loss—possibly millions of dollars per year over each of 50 years or so. But, nonetheless, restoration of that resource—not pounds of fish flesh—is the public sector manager's objective. The manager may suggest that short-sighted, profit-motivated, special-interest-group arguments, such as that of attributing restoration effort costs to pounds of fish flesh, is the line of reasoning that resulted in destruction or loss of resources in the first place. The public sector manager has a strong argument.

Sometimes Financial Analyses Do Not Miss the Point

Is there, then, to be no financial accountability or measure of efficiency in the public sector? In the private sector, there is a constant minimum threshold and a continuing pressure for efficiency in the forms of solvency and profit. The argument that public domain objectives can be of such a nature that the means to achieve them are not suited for economic evaluation can be so appealing that it may seem as if public managers do not want to address efficiency.

However, most public domain managers would agree that once an objective has been set, that objective should be achieved as effectively and efficiently as possible. Although the public manager is not in cost-cutting competition with a fish market supplier, neither is the manager free of accountability for effectiveness and efficiency. The fewer options and increased scrutiny associated with the public versus the private domain do not mean that the costs associated with accomplishing each objective should not be evaluated. *Profit* is simply an accounting term representing the surplus that remains after all costs are paid; it is the return to management (after costs are paid) for adding value through a production process. An argument that public processes cannot be financially evaluated in a manner similar to that of profit-making processes, is usually a copout. The costs and returns, the inputs and outputs, can all be evaluated and compared with like, or past, or expected process costs, returns, and so on. If in a public sector operation there is a surplus, there is something comparable to a profit; in public domain terms, the operation is *under budget*. The terminology may not be important, but the measure of efficiency is. Rather than avoiding the task of measuring success in economic terms, public sector managers need to consider the challenge of doing so in a meaningful way.

SUMMARY

Management is the planned, wise use of resources, including human, capital, equipment, and materials, to accomplish goals. Management functions have been categorized as planning, organizing, staffing, directing, and controlling. The manager is responsible for the results, good or bad, of the organization or business. Effective and highly productive managers are often "tuned in" to people, and ask for help and input from subordinates as well as from superiors.

The scope of aquaculture management covers the technical aspects of culture, such as the determination of water use and quality, the determination of what and how to produce, and the identification and solving of production and disease problems. But management also includes personnel aspects (staffing, training, motivating, and controlling), financial responsibilities, and accounting.

Management by objectives (MBO) is a bottom-line method in which the manager and employee set the standards or expectations for satisfactory job performance. Financial analyses of public (government) operations can be misleading unless the true objectives are adequately defined. Although public sector managers operate under restrictions, they should tie their operations to some measure of efficiency.

REFERENCES AND RECOMMENDED READINGS

Bayton, J. A. and R. L. Chapman. 1972. Transformation of Scientists and Engineers into Managers. Washington, DC: National Aeronautics and Space Administration.

Drucker, P. F. 1978. An Introductory View of Management. New York: Harper & Row, Publishers, Inc.

Frey, R. 1986. Rights, interests, desires, and beliefs. *In* People, Penguins, and Plastic Trees: Basic Issues in Environmental Ethics, ed. Donald Van De Veer and Christine Pierce, pp. 40–46. Belmont, CA: Wadsworth Publishing Company.

Fulmer, R. M. 1974. The New Management. New York: Macmillan Publishing Company, Inc.

Hammaker, P. M. and L. T. Rader. 1977. Plain Talk to Young Executives. Homewood, IL: Richard D. Irwin, Inc.

Kay, D. C., T. L. Brown, and D. J. Allee. 1987. The economic benefits of the restoration of Atlantic salmon to New England rivers. Human Dimensions Research Unit, Cornell University Series 87-6.

Likert, R. 1961. New Patterns of Management. New York: McGraw-Hill Book Company.

Likert, R. 1967. The Human Organization. New York: McGraw-Hill Book Company.

Morrisey, G. 1970. Management by Objective and Result. Reading, MA: Addison-Wesley Publishing Company.

Osburn, D. D. and K. C. Schneeberger. 1983. Modern Aquacultural Management (Second Edition). Reston, VA: Reston Publishing Company, Inc.

Weihrich, H. 1973. A study of the integration of management by objectives with key managerial activities and the relationship to selected effectiveness measures. Ph.D. dissertation. Los Angeles, CA: University of California.

Yukl, G. 1981. Leadership in Organizations. Englewood Cliffs, NJ: Prentice-Hall, Inc.

Chapter 3

Approaches to People
(Using Human Resources)

FOCUS:	People are the key to productivity, and productivity reflects the manager's effectiveness.
HIGHLIGHTS:	• Contingency theory
	• Importance of people orientation
	• Maslow's hierarchy
	• Active listening
	• Writing a performance standard
	• Efficiency and effectiveness
	• A checklist to rate yourself
	• Five steps to improving productivity
ORGANIZATION:	Contingency Theory
	Management Skills
	Leading and Motivating
	The Motivating (Maslow) Needs Hierarchy
	Communicating and Active Listening
	Using Performance Plans and
	Examples of a Performance Standard
	Productivity
	What Is Productivity?
	The Manager's Mandate and Enhancing
	Effectiveness
	Beginning
	Summary

CONTINGENCY THEORY

The contingency approach to organizational behavior is based on the tenet that it is difficult and at times incorrect to offer simple general principles to predict behavior within an organization. An extreme interdependence among individuals and situations influences or determines employee behavior, and human behavior is complex to begin with. Therefore the most effective approach to use with one employee or another, in one situation or another, usually cannot be standardized; rather, it is contingent upon many at times complex factors. Nonetheless there are some guidelines to follow in dealing with people.

The first management principle (Hammaker and Rader 1977) is to understand those you manage through a knowledge of and feeling for people. Good managers are people oriented, and to be successful, the manager must be able to select people and help them develop; deal fairly and persuasively with employees; set high standards; insist on achievement of the planned and required results; and reward superior performance and penalize suboptimal performance.

MANAGEMENT SKILLS

Leading and Motivating

Leadership is the ability to get others to do what you want, but its essence is the personal ability to influence, not the power to require. Hammaker and Rader (1977) offer a number of observations. Leadership is not bestowed, it must be earned. A leader has an interesting but not an easy life, and often must struggle to achieve exceptional clarity of purpose. A leader displays competence, has great assurance and confidence, and attracts followers. A good leader is a good teacher and has a clear set of beliefs, values, and goals. A good leader doesn't panic; when decisions must be reversed, the leader provides sufficient explanation to help followers accept the change and maintain confidence. A leader has courage.

The Motivating (Maslow) Needs Hierarchy

Maslow (1954) postulated that all people share five similar needs, which can be arranged in a hierarchy from fundamental body needs to self-fulfillment. Researchers have added "independence" to make six categories.

1. Basic: physiological life or body requirements for air, water, food, rest, protection, and shelter.
2. Safety or security: safeguarding against danger or deprivation.
3. Social: desire to be a part of a group; association; belonging; sharing and exchanging help; friendship and love.
4. Esteem: respect, achievement, competence, confidence, and so on.
5. Independence: working toward personal goals, authority, and freedom.
6. Self-fulfillment: attaining all that one can or can be, learning, growth and development, and self-actualization.

Unsatisfied needs act as motivators or stimuli and provide direction for actions. Unfulfilled needs define values and behavior. Basic needs must be satisfied first, and until they are, the higher needs are unimportant. Once basic needs are reasonably satisfied, security becomes important, and so on. A notable fact is that once a need is satisfied, it no longer acts as a motivator. There are many persons, however, whose needs at one level are apparently never satiated. Mitchell (1983) presented a thorough and much more involved explanation of the values, drives and needs among American social groups.

Communicating and Active Listening

Verbal communication is important, but nonverbal communication is the power component in building or destroying relations. In the presence of others we have a choice in sending verbal messages, but we have no choice except to send nonverbal messages. Since we must behave in some way, others are influenced by their perception of our behavior. For example, as our behavior is interpreted as a signal for legitimate anger by others, their response comes back to us. Once a cycle of behave, interpret, respond, behave, interpret, and so on is established, it serves no purpose to analyze the cycle to affix blame. If the purpose of analysis is to identify ways in which both persons might respond differently (right now), analysis may lead to improved relations. This may mean that the behavior or perceptions of either party may change, but only if there is a willingness to change.

McCormack (1984) asserted that knowledge of the ins and outs of every-day business life is largely a self-learning process; knowing the experience of others, however, may make the learning shorter, easier, and less painful. His "street smart" advice is defined as applied people sense or the ability to make active, positive use of instincts, insights, and perceptions. McCormack believed that, since business demands innovation, dependence on conventional wisdom is the biggest problem in American business, and managing a company is a constant process of breaking out of antiquated systems and challenging conditioned reflexes. Although some MBAs (executives holding the master of business administration degree) adjust to the real world, McCormack's' assessment was that most MBAs he hired were either congenitally naive or victims of their business training. They were afflicted with a sort of "real-life learning disability," a failure to read or size up people and situations properly, and "an uncanny knack for forming the wrong perceptions." He thought it an error to assume that advanced degrees or high IQs equal business sense, and concluded that neither are substitutes for people sense or street smarts.

According to McCormack, what people say or do in the most innocent situations speaks volumes about their true nature. They may act in one way with subordinates, in another with their boss, and in yet another with people outside their organization, but their real selves can't change with the environment. Business and management situations come down to people situations, and the sooner one knows about the person, the more effective one can be. McCormack suggested that a surprising number of executives lack an awareness of the true nature of their people and of what goes on around them, perhaps because they are too busy listening to themselves or too involved in their own corporate presence to notice what others are really doing. Insight into a person's unchanging true nature can be invaluable (the concept supports the modus operandi of professional psychics and fortune tellers). Insight requires opening the senses, talking much less, and listening more. Almost any situation will be handled differently by someone who is listening and someone who isn't. The single most consistent piece of business advice McCormack received from respected business friends, including several company chairmen, was "learn to be a good listener." One said "watch your listen/talk ratio." Unconscious body language is part of nonverbal communication, as is style of dress, and that which goes "unsaid" or is avoided can be conspicuous and easily "heard." The eyes are the most revealing organ for observation, and communicate thoughts more than anything else.

Ego, according to McCormack, is why some things that shouldn't happen do, and others that should don't. In a company of 250 people there are 250 egos, each with a unique view of reality. Most successful businessmen are "one giant ego with a couple of arms and legs sticking out," but a big, assertive ego may indicate a low self-image. Rather than challenging another's ego, one should use the information to manage the person or situation more effectively. Nothing blocks one's insight more than one's own ego. Appendix II contains a list of suggested steps for learning to read people.

Using Performance Plans and Standards

An essential part of the MBO system is involving each employee in developing the objectives and setting the standards for the job. Each employee should have several (three to seven) critical elements that constitute the essence of the job. Each element should have subelements that are the standards for the work to be accomplished. The standards should have several functions or characteristics:

1. Act like a job contract between the employee and the firm.
2. Address the most important aspects of the employee's duties and responsibilities.
3. Guide the employee to accomplish tasks that are most important to the position, as defined by the supervisor.
4. Be directed at specific tasks or requirements.
5. Act as units of measure (as do the standards maintained by the National Bureau of Standards), so that the employee's performance can be compared to the standard to determine if the performance measures up.
6. Be brief, clear, concise, believable, and easily understood.
7. Enhance employee–supervisor communication.

8. Encourage productivity and efficiency.
9. Enable employee and evaluator to agree on whether the accomplishments failed to meet, met, or exceeded the standards.
10. Be taken seriously by both supervisor and employee.

Standards should not have several other characteristics:

1. Be nebulous, tricky, confusing, or unmeasurable.
2. Include overemphasis on doing the boss's job or on completion of relatively minor administrative tasks.
3. Include unimportant or inappropriate tasks—especially those that are not the primary responsibility of the employee.
4. Be difficult to understand.
5. Be difficult to use as a unit of measure.
6. Discourage productivity.
7. Cause mistrust.
8. Mislead the employee about the importance of various job requirements.
9. Be a waste or abuse of time.
10. Be used if they are not used seriously.

Examples

Developing and writing MBO performance standards is, at first, a difficult and time-consuming task. It requires the supervisor or manager to think through the job, over the entire rating period or planning year, and tell the employee what is expected. There is no vague advice to "take your best shot, and I'll tell you someday if I liked it, or if you were on the right track." If the manager can state the need, in terms of results, and develop a standard with the employee, that standard will keep working for the manager, day by day, week by week, month by month. The following are examples, roughly in order of improving quality, of different forms of written performance standards. Every standard, for every employee, in every organization can be unique. The assumption behind these examples is that an employee is assigned a research work unit, and that the research findings must be reported. (A research product example is used because researchers often argue that their work cannot be measured or graded.) Might one expect the quality and usefulness of the employee's response to follow loosely the quality of the written standard? Consider the response you would expect, or give, in order to measure up to the following standards.

1. Ensure that a report is written for each work unit.
2. Submit work unit reports of a quality acceptable to your supervisor as requested.

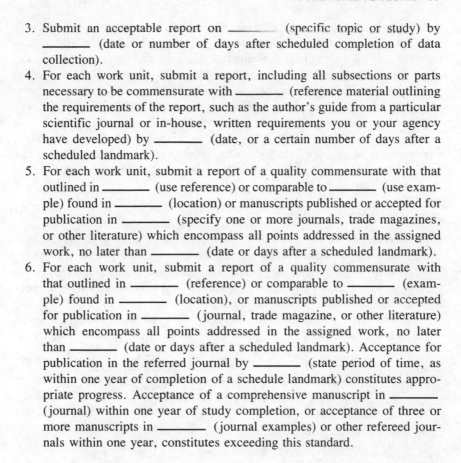

3. Submit an acceptable report on _____ (specific topic or study) by _____ (date or number of days after scheduled completion of data collection).

4. For each work unit, submit a report, including all subsections or parts necessary to be commensurate with _____ (reference material outlining the requirements of the report, such as the author's guide from a particular scientific journal or in-house, written requirements you or your agency have developed) by _____ (date, or a certain number of days after a scheduled landmark).

5. For each work unit, submit a report of a quality commensurate with that outlined in _____ (use reference) or comparable to _____ (use example) found in _____ (location) or manuscripts published or accepted for publication in _____ (specify one or more journals, trade magazines, or other literature) which encompass all points addressed in the assigned work, no later than _____ (date or days after a scheduled landmark).

6. For each work unit, submit a report of a quality commensurate with that outlined in _____ (reference) or comparable to _____ (example) found in _____ (location), or manuscripts published or accepted for publication in _____ (journal, trade magazine, or other literature) which encompass all points addressed in the assigned work, no later than _____ (date or days after a scheduled landmark). Acceptance for publication in the referred journal by _____ (state period of time, as within one year of completion of a schedule landmark) constitutes appropriate progress. Acceptance of a comprehensive manuscript in _____ (journal) within one year of study completion, or acceptance of three or more manuscripts in _____ (journal examples) or other refereed journals within one year, constitutes exceeding this standard.

Would you have given the same response to versions 1 and 6? Can the employee fail to meet version 1? In version 6 the supervisor is asking for and expecting to receive a well-defined product, and the employee knows exactly what is expected. The employee also knows exactly how to exceed the standard. Preparing well-defined standards involves more work, more planning, and more early communication. The effort should result in no surprises at the evaluation; that is, both the employee and the supervisor will know the level of performance without having to rely on judgment or personal persuasion.

It becomes evident that words such as *ensure* and *acceptable* are almost assuredly unacceptable in the clear enunciation of standards. The term *as requested* is difficult to use as a unit of measure. Standards should be based on things that are tangible. (A pencil is tangible, but my thoughts and opinions about an acceptable pencil are not. Similarly, a document, or reference to a document available to the employee is tangible.) The employee should be able to clearly fail to meet, meet, or exceed the standard in time, quality, and quantity.

It is not important to develop all three (time, quality, and quantity) within each standard as long as the employee and the evaluator understand from the written standard just what constitutes inadequate, adequate, and superior performance.

Often managers find, in retrospect, that early in the use of a new MBO system, many standards are not standards at all but are simply directives. Some indication of a standard for measurement of adequate performance is needed within each subelement. Specific responses should be stated if the supervisor or evaluator expects the employee to provide such responses. Failure to tell the employee the desired response defeats the purpose of the standard. The written standard should be the desired or "standard" response, so that the employee can strive to achieve or surpass that response.

PRODUCTIVITY

High productivity is the result of good management, and good management begins with the manager. Many executives find that significant and lasting improvements in productivity can result from a thoughtful and deliberate process involving each employee in an organization. A quick fix may have a quick payoff, but its effects are generally short-lived.

A manager should spend part of each day or week improving the effectiveness and efficiency of the organization. Periodically the manager needs to:

Identify and redefine the mission, needs, issues, and problems.
Collect information on how the operation is running.
Introduce change—try new ways of doing things.
Measure and analyze the results.
Make systemic or permanent changes.
Evaluate and continue to monitor.

Although there are some rules of thumb for improving productivity, doing so is still an art, and a manager will often get faster results by adopting/adapting someone else's ideas than by simply following theory. To maximize management skills, one needs to refer to a variety of resource materials and to talk with other managers. Much depends on the manager's attitude and the organization's climate for change. Everything depends on you and how you operate as a manager in your organization. With that in mind, the following discussion contains some theory, some practical approaches to consider, and some examples that may give insight into specific problems.

What Is Productivity?

Broadly defined, *productivity* is the measure of individual or organizational performance. It includes not only efficiency, or the ratio of outputs to inputs; but also effectiveness, the extent to which the output satisfies program objectives.

Productivity is concerned with both doing things right and doing the right things. It involves not just the amount of a product produced or service delivered, and at what cost, but also quality, timeliness, and responsiveness. Rearing 20% more fish doesn't make sense if mortality triples or the shipping operation can't keep up with the increase. A balance is required for improving operations.

$$\text{Efficiency} = \frac{\text{output}}{\text{input}}$$

Productivity = doing the right things right

An increase in productivity may include (1) increasing efficiency, (2) increasing usefulness and effectiveness, (3) increasing responsiveness to demand, (4) decreasing costs, and (5) decreasing production and delivery time.

The Manager's Mandate and Enhancing Effectiveness

Ours is a world of fixed and increasingly scarce resources; most managers are intimately familiar with the reality of smaller budgets. Similarly, most feel with growing urgency the demand for greater productivity: more product, higher quality, and more timely delivery, all for the same or lower costs. Today's manager deals with declining means and rising demands.

The manager is not simply an overseer of a process, but plays the pivotal, central role in production. The manager's strengths and weaknesses, strokes of genius and embarrassing foibles, are amplified by the organization. Thus, the true assessment of a manager's performance is through the measure of organizational accomplishments. Here are five steps you can take to increase management effectiveness.

1. Make goals and purposes clear from the start. Explain what you intend to do and why. You may not be entirely sure yourself, but explain your general strategy, initial impressions, timetable, and who you will ask for help. Make sure that all employees understand by putting your intentions in writing and, if possible, by explaining in person.

   The goal might be to improve a spawning process, eliminate a production backlog, speed up or reduce paper work, reduce staff by attrition, or change the output or impact. Be realistic about goals, and make sure subordinates know that they will have an opportunity to participate in the process. Even if you are given a predetermined mission or goal, your staff should be allowed to help plan how you and they will achieve it. Write down not only the goal but also the reasons for achieving it and the methods for doing so.

2. Rate yourself as a manager. Be honest. No manager is perfect, and one place to start looking at productivity is in your management performance.

You generally manage money, materials, space, equipment, and people. This discussion focuses on people.

In managing people you have an important, often overlooked, resource—your ingenuity, that is, everything you have come to know about people, such as what it takes to get the job done, what works and what doesn't, and the most appropriate and comfortable management style, given your personality. The people-related skills and knowledge you bring to your job are the keys to improving productivity. Many styles may work, but you'll want to ensure that your style gets the results you want. You might rate yourself as to the degree to which you:

a. Provide clear directions and routine feedback.
b. Give employees work that uses their skills and challenges their ability and intelligence.
c. Ensure that work crews are harmonious.
d. Give workers a voice in decisions that affect them.
e. Provide opportunities for advancement.
f. Create an environment that workers feel part of.

See Table 3-1 for a quick checklist to help rate yourself.

3. Involve people in looking for opportunities. Very likely you have some ideas about where and how your operations might be improved. Perhaps they involve training a few employees in job skills, simplifying a form, reorganizing your staff, or improving the facilities. Talk these over, informally at first, with your employees. Informal conversations often give a quick insight into their feelings. Be open; don't merely present your views and ask for approval. Use your ideas to prompt employees to think about productivity. Discuss the benefits of improved productivity for the organization and them. Here are some questions you might ask:

Where do opportunities lie for improvements?

What are the obstacles we need to overcome?

What tasks can we do more efficiently?

What might we stop doing?

Which of our procedures need to be shortened, simplified, or eliminated?

What are employees unhappy about?

How can we use performance appraisals to improve productivity?

What can I do personally to improve productivity?

Productivity has been a negative word to many employees, and some still suspect that it means management pressure. But many now accept the idea that a fairly administered productivity program can help both management and labor if both groups jointly identify and resolve productivity problems. Continuing dialogue between managers and employees, rather than stop-and-start consultation, encourages a positive attitude toward increasing productivity. Such dialogues are often best served by management committees or productivity councils.

One way to identify and solve work-related problems is to establish a *quality circle*, a group of 5–10 employees who voluntarily meet to solve problems. If preparation and training are careful and management is willing to permit employees to participate, quality circles can help increase productivity. Get the crew together and talk things over as a team. Once you have identified an area in need of improvement and considered ways to achieve that improvement, consider also the obstacles you might encounter and the means of getting around them. Neither management nor labor can make lasting improvements alone; productivity should be neutral ground in labor–management relations.

4. Analyze and measure both before and after you make changes. Many professionals resent the suggestion that their output can be measured or graded. It isn't easy to do. Here are some steps you might consider in setting up your own measurement system.

   a. Analyze your objectives and problems, along with opportunities for improvement.

      Even if your organization has well-established functional statements and procedures, take a fresh look at its entire mission and current operation. Your analysis should show how things really work; it should not just mirror your preconceptions or duplicate existing documents. Base your analysis on discussions with employees, but be sure that you have diagnosed and fully understood the problems and the opportunities to improve. Think through major options in terms of advantages, disadvantages and costs. Ask advice of those who know something about the options.

   b. Decide what ought to be measured.

      Activities that at first appear too complex or nebulous for adequate measurement are more manageable when broken into parts. Determine what part of the process can be controlled, and work within those boundaries. Distinguish between quantifiable and nonquantifiable products; that is, at the end of a given time, do you have a product in hand or simply time spent in an activity, with no tangible output to be evaluated?

   c. Define the output measures.

      Some of your measures should reflect quality changes. In a fish production unit where the measure of output is pounds of fish, changing feeding procedures to provide more uniform fish size and conditions is a quality change. If the output index is not adjusted to show uniformity, it will not reflect the change in quality.

   d. Determine the input measures.

      The simplest input is labor, in years or hours. This measure is adequate if personnel is the major cost, but if you also use large amounts of fish food, other materials, and equipment, you will need to look at the relations between labor and these other costs.

e. Establish the collection system.

Study existing accounting and management information systems to see if the data you need are available. If not, make sure that the system you create gives the information you need on time, accurately, and as simply as possible. Begin with broad measures; after you get a "feel" for the measurement system, refine it to compare data over time or to compare it with other units doing similar work.

f. Analyze the validity and usefulness of the system.

Once the data collection system is in place, continue to evaluate its validity and usefulness. Determine if the data adequately represent current activities and if they provide useful information for decision making. It is important that you and your staff continually review the data system and refine it when necessary.

Managers worry about the results of measuring productivity for three reasons: (1) measures have often been inadequate; (2) the data have sometimes been used improperly in making decisions; and (3) some believe that productivity increases will be used as reasons to reduce their budget. To avoid poor use of measures, discuss with all managers the potential effects of improving productivity.

5. Choose opportunities. Once you have analyzed your work, initiated a measuring system, and identified problems, look at approaches you might use.

a. *People approaches*. Your most important resource is employees. But few employees work at their full potential. You set the standards, provide the direction, authorize training and appraise their performance, so it's your responsibility to ensure that they are both efficient and effective. Goals for performance must be specific and precisely stated; they must be of suitable difficulty; and they must be accepted by the employee. You need to work with the employee in setting each goal, and you must be sure that the employee "owns" the goals. You also need to ask some hard questions of employees and of yourself. For example:

(1) How much time does your staff spend on nonessential work, such as assisting with work that others are accountable for?

(2) In response to demands for more production, do you tend to recruit more employees, or do you try to increase your staff's capability? Money spent on more staff might be better spent on upgrading people or equipment, or redesigning procedures or work styles.

(3) When do the heaviest demands on staff time occur? Part of each day or week? Seasonally?

(4) If you have a good employee in what has become a mediocre job, you will soon have a mediocre employee. Do you delegate max-

imum authority, responsibility, and accountability? Some managers underuse their staff because they underrate them. You might increase effectiveness by decentralizing authority and holding lower-level employees more accountable for their performance. As a result, employees may find their jobs more challenging and rewarding.

(5) Do you train people adequately to solve real problems, or is training a reward? Do you spend time counseling employees to identify and correct weaknesses?

(6) Do you properly reward good performance and correct poor performance? Do you believe that an employee's salary is sufficient reward for good work, or do you try to understand what will motivate continued good work or perhaps outstanding work? One key is a performance appraisal system that requires you to identify and explain the critical elements of each employee's job.

b. *Procedural approaches.* Most large organizations have many procedures. One way to analyze procedures is through activity flow charts. Flow charts should show how things really work, not how someone believes they work. You may find that a report or contract requires more approvals than necessary, or that employees don't understand the total process they're part of but focus only on their tasks. Involve workers as well as supervisors. Those who do the work usually know the most about it.

c. *Environmental approaches.* Physical environment strongly affects behavior, and the work environment is often only a small part of the total cost. Improving the work space can be an inexpensive way to increase productivity.

The environment can be used as a tool that enhances work. The shape of the crew room, shop or office, its spaces, placement of equipment, light, color, and sound can all support (or hinder) its occupants and their work. Productivity is affected by personal space, territoriality, privacy, noise, lighting, availability of shower and locker rooms, a clean and comfortable break or lunch room, and other factors.

Corporations, including McDonald's, Weyerhauser, and Reader's Digest, have analyzed and revamped work areas with measured results of increased productivity. The best payoff results from a heavy up-front investment in time for evaluation and planning, and then the redesign implementation, including purchase of the right kind of (often more expensive) equipment.

d. *Capital investments: budget considerations.* Enhancing productivity by investing in new technologies has been the major reason for fast-growing productivity in factories and farms, but not in government, except for computers and word processing equipment. The lack of government investment has had three causes: (1) major spending on equipment usually requires a long lead time and detailed justification; (2) many managers do not have enough information about equipment to understand its implications for reducing costs and improving operations; and (3) managers are reluctant to make investments because they risk their reputations.

Nonetheless, relatively small capital investments can and do provide a payoff. Managers can err here, however, just as easily as they can in managing personnel; there are abundant examples of capital investments that did not improve productivity or reduce operating costs. It is essential to solicit staff or worker input to manage equipment efficiently. The preinvestment analysis must be complete and the justification correct.

Beginning

You may wonder what to do first. Many managers find the problem of improving productivity so big that they make up a long-range plan and then gradually forget it, or make a few changes and then slide back into the old routine. An important thing to keep in mind is that improving productivity is a long-term process of development—of yourself, your management, your employees, your entire organization. It takes continuous evaluation and analysis, probing for improvements, and reinforcement of all persons involved. It is a combination of persistence and creativity.

Table 3-1. Checklist to help rate yourself as a manager. Too many "no" answers probably mean danger. Too many "yes" answers may mean that you are kidding yourself.

	YES	NO
1. All my employees fully understand our goals and objectives.	( )	( )
2. All my employees are capable of doing the task assigned to them.	( )	( )
3. All my employees understand their standards of performance and feel that they are realistic.	( )	( )
4. I can measure the efficiency and effectiveness of my organization's work.	( )	( )
5. I encourage innovation and teamwork.	( )	( )
6. I improve working conditions whenever possible.	( )	( )
7. My employees have a sense of ownership of their program.	( )	( )
8. If employees are not working up to standards, I quickly find out why.	( )	( )
9. I reward good or outstanding work promptly.	( )	( )
10. I have a plan to evaluate my organization's productivity periodically and to improve it.	( )	( )

Recognize and accept the importance of improving productivity—not much will happen unless you are committed. You must believe that something can be done, and you must be enthusiastic.

Use your existing authority. Improve the use of awards and incentives, for example, by allocating a percentage of your budget for rewarding and stimulating improved work.

Establish working alliances and downplay adversarial relations. Get to know managers of similar programs, and become acquainted with budget and personnel managers.

Continue to clarify priorities, goals, objectives, and measurements.

As you proceed, communicate with all those involved, explaining what you are doing and why and their role; continue to keep all of them informed of progress, change, results, and setbacks.

The five Steps to improving productivity can be summarized as follows:

1. Identify the purpose and timetable: Clearly identify the goals for all those involved. Establish realistic timetables.
2. Rate yourself as a manager: Use and sharpen your skills; use your full authority as a manager.
3. Involve people: Involve all employees and top management. Seek their support; keep them informed; be sensitive to their needs; let them know of successes and failures; and keep their support alive.
4. Measure and analyze: Set up a measurement system; keep it simple at first. Set up a system to analyze the information, and use your experience and judgment to interpret the data.
5. Choose opportunities: Identify functions, units, and procedures that are ripe for productivity improvement. Locate models in other organizations. Implement them on a small scale first.

SUMMARY

Good managers make a point of understanding the people they manage. The Maslow needs hierarchy suggests that unfulfilled needs define values and behavior; an understanding of the motivating power of needs can benefit the manager as well as the employee. Active listening is a powerful tool that can enhance communication and effectiveness. Performance plans and standards are developed around management objectives by the employee and supervisor. The plans provide constant direction and thus aid in maintaining control. The plans also act as a job contract, provide a systematic method to enhance communication, and help reduce the judgment role of supervisors in evaluation.

Productivity is a combination of efficiency and effectiveness. It is central to the purpose of management and constitutes an area that most managers should systematically address and improve. Five steps to improving productivity are:

Identify your purpose and timetable.

Rate yourself as a manager.

Involve people.

Measure and analyze.

Choose opportunities for productivity improvement.

REFERENCES AND RECOMMENDED READINGS

Bliss, E. C. 1976. Getting Things Done: The ABC's of Time Management. New York: Charles Scribner's Sons.

Fast, J. 1971. Body Language. New York: Pocket Books.

Geldard, F. A. 1968. Body English. Psychology Today 12:42–47.

Goffman, E. 1959. The Presentation of Self in Everyday Life. Garden City, NY: Doubleday Books.

Hammaker, P. M. and L. T. Rader. 1977. Plain Talk to Young Executives. Homewood, IL: Richard D. Irwin, Inc.

Longfellow, L. E. 1970. Body talk–a game. Psychology Today 10:45.

Maslow, A. 1954. Motivation and Personality. New York: Harper & Brothers, Inc.

McCormack, M. H. 1984. What They Don't Teach You at the Harvard Business School. New York: Bantam Books.

Mehrabian, A. 1968. Communication without words. Psychology Today 9:53–55.

Mitchell, A. 1983. The Nine American Lifestyles. New York: Warner Books.

Nierenberg, G. I. and H. H. Calero. 1973. How to Read a Person Like a Book. New York: Pocket Books.

Rosenthal, R., et al. 1974. Body talk and tone of voice: the language without words. Psychology Today 9:64–68.

United States Office of Personnel Management Manager's Handbook. 1981. Based on United States Office of Personnel Management Manager's Guide for Improving Productivity, #WP-1. Washington, DC: U.S. Office of Personnel Management.

Vecchio, R. P. 1988. Organizational Behavior. Chicago: The Dryden Press.

Wiener, N. and A. Mehrabian. 1968. Language within Language: Immediacy, a Channel in Verbal Communication. New York: Appleton-Century-Crofts.

Chapter 4

Marketing

INTRODUCTION

Marketing is all the business activity involved in the transfer of a product to a consumer. It is not simply salesmanship or advertising, although it includes those functions, but a varied and complex field of endeavor and study. Marketing permeates our social and work lives. Employees market their abilities (potential products), and organizations—from churches to government agencies—market their offerings. Where there is something of value or utility to be used or transferred, marketing is taking place or marketing opportunities exist. Commercial marketing has cultural implications and generally involves processing and distribution, with brokers in between. Figure 4-1 gives some insight into the complexity and the interactions commonly involved in the marketing of an established product, such as a fish fillet.

The word *market* can refer to a place, a product, a time, or a level (wholesale or retail) within the industry. A market has characteristics of movement, or a series of actions and events, and coordination, as the series must occur in a sequence. The field of marketing offers prime management opportunities, and

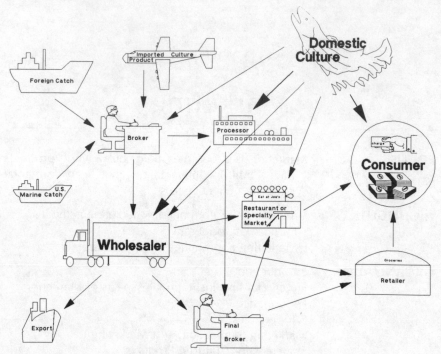

Figure 4.1. Interaction of typical commercial marketing activities in fisheries.

extensive literature, such as Britt (1973), is available on the scope as well as on detailed aspects of marketing management.

FUNCTIONAL AND INSTITUTIONAL ASPECTS OF MARKETING

Marketing is often presented from either a functional or an institutional point of view. The functional approach relates to the job or to the way the market operates, and a study of each of the four functions is useful for cost evaluation.

The exchange or title transfer function involves the buying, merchandising, and differentiating of the size, process, quality, and so on.

The physical function concerns actual handling, movement, storage, and processing of the harvested product or solving the what, where, and when problems of product movement to the consumer.

The facilitating function involves three subfunctions: standardization, financing, and risk bearing. Standardization of the product enhances the exchange and physical functions by simplifying markets and allowing mass marketing. Financing, often done at the producer's level, is required during the time delay between production and consumption, when capital is tied up. Risk bearing involves taking responsibility for physical aspects, such as product deterioration (spoiling), and market aspects, such as value (price) change during the in-market period.

The intelligence function encompasses the collection, interpretation, and assimilation of data to improve the market (technically this is part of facilitating).

The combination of certain marketing steps with production, usually to reduce costs, is referred to as *vertical integration*. The four functions add value as they transform a product into a consumable item. When the functional value that is added is greater than the added cost, the consumer benefits and the producer realizes a competitive advantage.

The institutional approach to marketing involves the description of the jobs or of the individuals involved.

A merchant middleman, part of an institution or an organized group, takes title to the product and performs a direct market function as a wholesaler or retailer.

An agent middleman is either a broker, who does not take title, or a commissioner, who takes possession and then resells.

A speculative middleman is one who takes title with the specific goal of selling high and fast.

A processor takes title and changes the form, sometimes referred to as a *value-added step*, of the product.

A facilitative organizer, such as a trade association, provides a marketplace (but does not participate in the market) and may provide information, govern ethics, or set standards.

PERFORMANCE

Market performance can be critically important to the culture system manager. *Presumptive concerns* include product quality, packaging, labeling, and product measurement (as by weight or standard). The market performance itself can be measured through trend or time series comparisons, price stability (of the product) and income (of the producer), percentage of the consumer's budget (total or grocery income, etc.) spent on the product, market margin or producer-to-retailer price spread, and the profit margin at each market level.

Another measure of market performance is *operational efficiency*, or the change in the cost of marketing (without altering the product) that reflects labor output per labor hour used. *Improper efficiencies* refer to such things as the inability of the consumer to purchase a desired product, low product quality, or a failure to coordinate buying and selling activities in the marketing process. Equitable pricing is conducive to the smooth flow of the product through the market. Poor pricing can result in improper allocation of products, overconsumption of scarce products, high fluctuation of profits (and losses), and externalities. An *externality* is a change in the cost or operation of one fish producer caused by another producer, and a market imperfection is referred to as a *pecuniary externality*.

COOPERATIVES

Cooperatives sometimes play important roles in the successful development of aquaculture industries. A cooperative is an association of individuals or businesses joined together to achieve a gain; the members are the patrons and the owners, in contrast to the investor-owned, public patronage corporation. The most common problem of a cooperative is management, and the key to success is the manager.

The objective of a cooperative is to maximize the return per unit of cost for its members as a group. Through the structuring of an institutional process, its members can take group action to compete more effectively. Significant operational costs are incurred; for instance, someone is usually paid to run the organization, and the cooperative must limit nonmember business. But the cooperative benefits its members or users in ways such as reducing costs or obtaining better market prices. Cooperatives are especially advantageous in developing countries.

There are four general types of cooperatives. A marketing coop replaces a function or institution in the marketing chain. A service coop provides information, materials, credit, electricity, hospital care, or other services or service opportunities for its members. A processing coop changes the form of the product; for instance, it may kill, clean, package, and freeze fish. Finally, a producer (consumer) coop works toward cost reduction through such means as purchasing feed or netting in bulk.

Success depends not on uniqueness but on the business advantage gained. If the purposes or objectives are worthwhile, success depends on achieving them; achieving objectives is central to the purpose of management, and thus the manager becomes the key to success. There are unique problems associated with the management of cooperatives. There is a tendency for successful cooperatives to shift toward investor welfare (corporate structure) rather than patron welfare. The government therefore often limits nonmember patronage. The structure of a cooperative is voluntary, and that can be a problem. Some marketing experts believe that control of supply, and thus stability or control of price, is the single most important aspect of marketing success, and organized suppliers can exert control. Usually a coop cannot control the price directly. If it is large enough, it can attempt to regulate the supply, but that depends on the voluntary compliance of its members. Finally, community nonmembers often unite against the coop because nonmembers are sometimes adversely affected by coop competition.

MARKETING STUDIES AND TEST MARKETING

Marketing research is often neglected and, until recently, has been almost nonexistent in aquaculture industries. The topic is unfamiliar to most culture system managers, and few believe that marketing research can be accomplished inexpensively. In the 1980s, marketing surfaced as one of the most urgent needs for, or its lack a major impediment to, the growth potential of commercial aquaculture in the United States.

Marketing research is any planned, organized effort to gather facts and develop knowledge to improve market decisions. Commonly but unknowingly, it is accomplished routinely by many sales and marketing people—including culture system managers who generate sales. With a little preparation, one can develop a more formal, objective approach. The results of a formal approach can usually be used and compared through time, and other people can collect the needed information. Seat-of-the-pants hunches may get one by, but as growth proceeds, or when other interests or investors are involved, as in cooperatives, the establishment of an objective system to collect and analyze market facts quickly becomes essential.

In planning a marketing study, one must look at supply as well as at demand. Starting with the consumer, the population is defined in terms of the fishery products it consumes. Sampling may be required, and steps should be taken to make the effort valid. Statisticians refer to *bias* as the tendency for the sample value to deviate above or below the true value. An *unbiased statistic* is one in which high and low sample values tend to average out. *Reliability* or *precision* is the closeness of the statistic (sample value) to the true value. *Consistency* is the tendency of the sample value to become a better estimate of the true value as the sample size increases. Sampling schemes can be complex and confusing, and to be valid, samples must meet certain criteria. Therefore, one should consult (read and study) a sampling methods reference or consult a statistician, as well as study a reference on marketing research.

STRATEGIES FOR CULTURED PRODUCTS

Products are either intended for human consumption, directly or indirectly, or they are not. Some general guidelines and strategies can be applied to both categories. For food items, however, certain important advantages associated with health should not be overlooked.

Generally speaking, the customer may not be everything, but the customer is at least equivalent to the queen and king rolled into one. Ignore the customer's "royal commands," and your head will roll. A key to success is providing the customer with what he or she wants or prefers. Hence the value of marketing research discussed in the previous section.

Once customer preferences have been identified, marketing objectives must be set. The manager needs to know where the business should be going to have a reasonable chance of getting it there. The manager should use all resources available and consider brainstorming. There are no criticisms allowed and no bad ideas in a brainstorming session. The more unusual or unique the idea, the more likely it will elicit criticism, but also the more innovative and valuable it is likely to be. The manager should consider all possibilities, look at interests and strengths, and then set objectives. The objectives will then direct the manager to and through a marketing plan.

When developing objectives and a marketing plan, one should not be confined by tradition. New products, promotion, and advertising may be important, but the manager cannot afford to overlook the current product. Competition and marketing are conducted in ways other than by simply lowering the price. The manager must consider quality, availability, and enhanced appearance or packaging; and then segment the market, looking for new customers and increasing exposure. Finally, the manager should consider offering related services or products before changing the standard product.

The built-in advantages of fish and other aquatic products are numerous, but those related to health have the greatest potential marketing advantages. A good rule to follow might be: "When marketing fish, never market fish." You may be selling fish, but consider marketing something else. Most car manufacturers have not marketed transportation since . . . can anyone remember? They market luxury, comfort, status, attractiveness (sex?), independence, individuality, and other concepts.

Fish are said to be healthy. They contain high protein, low fat, low cholesterol, and few calories. Ocean fish have gained attention due to mounting evidence of the beneficial health effects attributed to omega-3 polyunsaturated fatty acids. But bioaccumulation of pollutants is a customer concern. Aquaculture products may not have the omega-3 advantages of ocean fish, but they do have the advantage of a known rearing environment, so that pollutants can be avoided. And the products can be marketed as offering health or dietary advantages (for weight control), as well as all the desirable ancillary attributes of attractiveness, youth, longevity, and vigor. (And unlike the characters in Aldous Huxley's *After Many a Summer Dies the Swan*, today's aspiring centurions don't have to eat raw viscera!)

The nonfood items also have some good built-in advantages. Since a 1983 University of Pennsylvania study indicated that aquarium viewing reduces blood pressure, pet store fish sales have risen—to $219.7 million in 1985. As for sport fish, some unrelated product marketing schemes can show the way, with ads featuring fishing along with catch phrases such as "It doesn't get any better than this" or simply "Gone fishing." Stress reduction, escape, relaxation, pleasure, and peace of mind are the marketable concepts.

Consider some don'ts. Don't overlook the homemaker, whose career involves managing the nutrition, the health, and the grocery budget for the family. But don't market to the antiquated, stereotype of the homemaker.

Don't confuse marketing with selling, so that you focus only on sales; if you overlook consumer preferences and geographical distribution patterns, you may be ignoring royalty. And finally, don't give your product to a broker based on a consignment sales agreement. Brokers work harder to sell something in which they have an investment.

SUMMARY

Marketing is a complex field that involves all the activities associated with the transfer of a product to a customer. The functional approach relates to the job of marketing, or the "how to" of the exchange, physical, facilitating, and intelligence functions of market operation. The institutional approach describes the persons involved. Market performance is defined, as it is measured, in a monetary term such as price stability or income, or in a percentage measure such as the consumer's budgets. Performance is also measured in efficiency as the cost of marketing per unit of effort.

Cooperatives are patron-owned associations for mutual gain and have unique management problems. Individuals, as well as cooperatives and large businesses, can feasibly conduct marketing studies that measure customer wants to identify opportunities. Marketing opportunities need to be weighed to develop an effective mix in much the same way as products and enterprises are mixed. The health aspects of fish and seafoods provide a built-in marketing advantage for aquatic products. The advertising aspect of marketing should suggest satisfaction of personal needs beyond food. Don't market fish; instead, market health, longevity, fitness, attractiveness, and so on.

REFERENCES AND RECOMMENDED READINGS

Anderson, L. G. 1977. The Economics of Fisheries Management. Baltimore: Johns Hopkins University Press.

Anon. 1987. Composition of Foods: Finfish and Shellfish. USDA Handbook 8–15. Washington, DC: U.S. Government Printing Office (stock no. 001-000-04497-4; $10).

Anon. 1987. Marketing excellence awards. Seafood Business 6(5):70–85.

Anon. 1987. State Fish and Seafood Supplies Directory. Albany, NY: New York State Department of State.

Barnett, J. 1987. Mississippi catfish and the chicken boys. Seafood Business 6(5):58–65.

Bell, F. W. 1978. Food from the Sea: Economics and Politics of Ocean Fisheries. Boulder, CO: West University Press.

Breen, G. E. and A. B. Blankenship. 1982. Do-It-Yourself Marketing Research. New York: McGraw-Hill Book Company.

Britt, S. H. (ed.) 1973. Marketing Manager's Handbook. Chicago: Dartnell Press.

Chaston, I. 1983. Marketing in Fisheries and Aquaculture. Farnham, Surrey, England: Fishing News Books Ltd.

Davies, R. 1981. A new look at the United Kingdom fish market. Fish Farmer 4(7):40–41.

Dyerburg, J. 1986. LinoLenate-derived polyunsaturated fatty acids and prevention of atherosclerosis. Nutrition Reviews 44:125–134.

Elliott, D. 1984. Fee fishing—California style (How to fill your lake with fishermen). Aquaculture Digest 9(7):6–10.

Food and Agriculture Organization of the United Nations. 1986. Development of an international trade in fishery products, 1960–1985. COFI:FT-I/86/2. Rome: FAO.

Frand, E. A. 1983. More marketing myopia. Industrial Research and Development 25(6):23.

Garza, D. A. 1985. Market structure and marketing potential for Kotzebue chum salmon. Marine Advisory Bulletin 16. Anchorage: University of Alaska.

Gasson, R. 1977. Farmer participation in cooperative activities. Sociologia Ruralis 17(1):107–123.

Globefish. 1987. Fisheries update 1986. Infofish Marketing Digest, 87(4).

Glude, J. B. 1983. Marketing and economics in relation to U.S. bivalve aquaculture. J. World Mariculture Society 14:576–586.

Hed, A. 1985. National agriculture export council building new markets. Aquaculture Magazine 11(4):28, 30–32.

Heller, W. P. and D. S. Sterritt. 1976. On the nature of externalities. In Theory and Measurement of Economic Externalities, ed. Steven A. Y. Lin. New York: Academic Press.

Israel, D. C. 1987. Comparative economic analysis of prawn hatcheries. SEAFDEC-Asian Aquaculture 9(2):3–4,12.

Jordon, S. 1987. Fish farmers cast for big market. Salmonid 11(3/4):27.

Kabir, M. and N. B. Ridler. 1984. The demand for Atlantic salmon in Canada. Canadian J. Agric. Economics 32:560–568.

Kabir, M. and N. B. Ridler. 1985. The demand for Atlantic salmon in Canada: A reply. Canadian J. Agric. Economics 33:247–249.

Kabir, M. and N. B. Ridler. 1985. The future of salmon farming in Atlantic Canada: A quantitative approach. Presented to the Annual Meeting of the Canadian Economics Association, May 29–31, 1985, Montreal, Canada.

Kotler, P. 1980. Marketing Management. Englewood Cliffs, NJ: Prentice-Hall, Inc.

Kotler, P. 1982. Marketing for Nonprofit Organizations. Englewood Cliffs, NJ: Prentice-Hall, Inc.

Kotler, P. 1984. Marketing Management: Analysis, Planning and Control. Englewood Cliffs, NJ: Prentice-Hall, Inc.

Lands, W. A. 1986. Fish and Human Health. Orlando, FL: Academic Press, Inc.

Leftwich, R. H. and R. D. Eckert. 1982. The Price System and Resource Allocation. Chicago: The Dryden Press.

Leonard, D. K. and D. R. Marshall, eds. 1982. Institutions of Rural Development for the Poor: Decentralization and Organizational Linkages. Berkeley: Institute for International Studies, University of California.

Liao, D. S. and T. I. J. Smith. 1981. Test marketing of freshwater shrimp, *Macrobrachium rosenbergii*, in South Carolina. Aquaculture 23:373–379.

Liao, D. S. and T. I. J. Smith. 1982. Marketing of cultured prawns, *Macrobrachium rosenbergii*, in South Carolina. In Proceedings of the 13th Annual Meeting of the World Mariculture Society, March 1–4, 1982, Charleston, SC, ed. James W. Avault, Jr., pp. 56–62. Baton Rouge, LA: Louisiana State University Press.

Lin, B. H. 1984. An econometric analysis of salmon markets. Paper presented at the Second Conference of the International Institute of Fisheries Economics and Trade, Cristchurch, New Zealand.

Lin, B. H. and N. A. Williams. 1985. The demand for Atlantic salmon in Canada: a comment. Canadian J. Agric. Economics. 33:243–246.

Lovshin, L. L. and R. Pretto. 1983. A strategy for the use of tilapia in rural Latin America: the Panamanian integrated approach. Technical Report. Auburn, AL: International Center for Aquaculture, Auburn University.

Mellor, J. W. and B. F. Johnston. 1984. The world food equation: Interrelations among development, employment, and food consumption. J. Economic Literature 22(3):531–574.

Melteff, B. R. 1983. Proceedings of the International Seafood Trade Conference, September 8–12, 1982, Anchorage, AK. Alaska Sea Grant Report No. 83–2.

Molnar, J. J., N. B. Schwartz, and L. L. Lovshin. 1985. Integrated Aquaculture. Auburn, AL: Auburn University.

Nettleton, J. A. 1987. Seafood and Health. Huntington, NY: Osprey Books.

New England Fisheries Development Foundation and National Fisheries Institute. 1984. Point of purchase resources. Seafood Business Report 3(1):43.

Pilkington, C. 1984. Marketing . . . not just for big business anymore. Salmonid 7(5):10–12.

Pilkington, C. 1984. Promotion and publicity. Salmonid 7(6):16–18,20.

Ridler, N. B. 1983. A preliminary economic evaluation of salmon aquaculture in the maritimes. Presented at the 12th Annual Conference of the Atlantic Canada Economic Association, Fredericton.

Seehafer, D. A. 1985. San Diego study emphasizes some basic marketing requirements. Aquaculture Magazine 11(4):23–26.

Shaw, S. and J. F. Muir. 1987. Salmon: Economics and Marketing. Portland, OR: Timber Press.

Stokes, R. L. 1982. The economics of salmon ranching. Land Economics 58(4):473–477.

Tsoa, E., W. E. Schrank, and N. Roy. 1982. U.S. demand for selected groundfish products, 1967–80. American. J. Agric. Economics. 64(3):483–89.

Waldrop, J. E. and J. G. Dillard. 1985. Economics (of channel catfish culture). In Channel Catfish Culture, ed. C. S. Tucker, pp. 621–645. Amsterdam: Elsevier Science Publishing Company.

Chapter 5

Life Cycles and Production Strategies

FOCUS:	Selection and mixing of enterprises are fundamental to the efficient use of resources.
HIGHLIGHTS:	• Concept of mix and the use of resources • Resources direct enterprise selection • Listing of typical enterprises • Operator preference is important for enterprise selection • High food conversion rates associated with some enterprises are misleading • Mix enterprises to maximize returns
ORGANIZATION:	Enterprises and the Concept of Mixing Energy Flow Summary

ENTERPRISES AND THE CONCEPT OF MIXING

New culture technologies are available; for example, a culturist can routinely induce or delay spawning of many cultured organisms. As an aquatic organism develops and changes form, it changes its requirements for nutrition, oxygen, space, temperature, water quality, and so on. The combination of technologies and changes in species with life cycle stages results in many possible combinations of specialization and enterprise. An *enterprise* is that which is sold, as well as the resources used to produce it.

The manager's job in setting objectives to reach organizational goals includes selecting, or at least recommending, the mixture of enterprises. The manager's challenge is to use all available resources, including time, when each resource is available. Considering the usually large seasonal differences in activities, enterprises should be complementary to maximize labor efficiency.

That is, in addition to the use of temporary or seasonal labor, and good planning and scheduling of full-time personnel, enterprises should be selected and combined in a way that allows an enterprise with a high resource use to "fill in" during the period of low resource use of another enterprise.

Suppliers of fish eggs, oyster spat, and postlarval shrimp are specialists within a specialty industry. The "seed" industry is well suited for specialists who can develop production, handling, and shipping techniques. Most of the problems in fish culture have historically centered on newly hatched fry. Once the eggs hatch and the fry swim up, begin to feed, grow, and reach a size that does not require nursing, the culture procedures become relatively routine and unchanging until the fish are harvested. Care of eggs and fry is labor intensive, but more than that, it is an artisan's job. A culturist who cares about the fry—who does not see cleaning as a chore but as an opportunity to improve the stock—and who worries about the health of the smallest organisms is rare. An organization managed by, or one that identifies and develops, such a person has the wherewithal to specialize in the production of seed and juveniles.

The possession or control of extensive water resources enables volume production of harvestable (edible or stockable) fish. Although quality can never be overlooked, quantity becomes the key in production and staffing. A manager does not need highly trained biologists or experienced and perceptive hatchhouse specialists to drive trucks and haul seines. The development, care, and selection of a broodstock, however, require technical expertise. The larger the operation, the more options in enterprises and staffing, but also the greater the chances of time and labor inefficiencies.

Although it may be feasible to produce a cold water, marine crustacean in Arizona, in any given location the array of the more profitable enterprises will be dictated by water quality and temperature. Thus, the choice of enterprises is often limited to those such as fresh- or saltwater, cold- or warmwater species, and so on. There are, however, myriad choices. In a well-populated area of a north-central state, a trout farmer might investigate such options as baitfish, crawfish, goldfish (in ponds), watercress, or even grow-out (finishing) of pike or walleye fingerlings. However, investigation of enterprises and optimization of the enterprise mixture do not require the use of new or additional species.

Whether the manager rears catfish, trout, or salmon, a variety of enterprises are usually involved. Most hatchery operators rear and keep broodstock, spawn and incubate eggs, nurse fry, produce juveniles, and raise at least one species to harvestable size. In addition to engaging in the four or five major enterprises, the operator may transport live fish, harvest ponds, or do some other related enterprise such as operating a fish-out (fee-fishing) pond, processing fish, or using part of the facility for wholesaling fish into or out of the area. Table 5-1 lists some enterprises commonly found in various aquaculture systems.

Table 5-1. Typical enterprises common to various types of aquaculture and various species

Brood production
Seed (egg, spat) production
Juvenile production
Adult or consumable (grow-out) production
Harvesting (custom)
Live transport
Public interactive enterprise (Fish-out ponds, roadside sale, information and training)
Specialty products
Culture support devices (net construction and repair, etc.)

If the operator transports fish or uses a tractor only infrequently, the high cost of owning the equipment must be recognized. For instance, is a $3,900 estimated annual ownership cost for a $20,000 tractor justified? Would it be cheaper to rent equipment when needed? If the money spent for the annual cost were placed in an interest-bearing account, the annual interest would represent the opportunity cost. The purpose of calculating the opportunity cost of all owned property becomes evident.

In addition to looking at equipment cost and return, the manager must also look closely at the returns to each enterprise. Enterprises that pay the highest returns should be maximized to the exclusion of those that yield the lowest returns. Maximizing returns introduces the question of just how to select enterprises. One of the most important criteria, which is often not identified for selection of enterprises for the small or family business, is operator preference. Operators are more likely to work well in preferred enterprises than in those they dislike. Choices must be limited, of course, to those enterprises that are economically feasible and potentially profitable. Here are seven criteria to use in selecting possible enterprises: (1) personal preference (of the owner-operator), (2) markets, (3) water type (salinity) and temperature, (4) water quality, (5) available capital, (6) skills and knowledge of the owner-operator, and (7) land and location suitability.

ENERGY FLOW

Energy for growth and metabolism is provided through food in a relation biologists refer to as a *food pyramid*. At each trophic level there is about a 6:1 reduction in biomass. In low-intensity forms of pond culture, food is provided by plankton that is supported through fertilization. Fertilizers may occur naturally, be wastes from human activities, or be purchased. In tidal areas food flows in with the tide, and in river systems allochthonous food is carried in from the land upstream.

In more intensive culture systems, essentially all food is provided by the operator. For fish the diet is usually in the form of a pellet. Food to flesh conversion rates, important to efficiency or profitability, are usually well below 2:1 (weight to weight) and sometimes below 1:1. The apparently high efficiencies, reaching or exceeding 100%, are misleading. The biological relations between trophic levels remain in effect but are confounded by the feed production process. In nature, the moisture content of food organisms exceeds 70% (sometimes 90%), whereas that of milled diets is 10% or less.

At a given temperature and feeding rate, fish length tends to increase in a fixed and linear fashion. Length is linear, but weight is weightier (geometrically) than it is linear. In general, fish weight increases as the cube of the length. The implication is that in manipulating growth rates, at variable costs such as temperature, excess food, and reduced intensification, it is much cheaper to increase the rates or the size when fish are small than when they are large.

Production strategies vary greatly to allow managers to take advantage of resources, as well as to provide the requirements of the cultured organisms. The variety of production strategies can increase opportunities for complementary mixing of enterprises.

SUMMARY

New technologies and the differences in culture requirements among the species and the life cycle stages of aquatic organisms provide the opportunity for an aquaculture manager to use resources efficiently through enterprise mixing. Resources dictate to some extent the feasibility of the enterprise. Operator preference can be very important in enterprise selection for the small business. The mix of enterprises should maximize returns to the resources consumed. The apparently high efficiency, or food to flesh conversion rate, of some pelleted diets can be explained by the moisture content of those diets. All input costs should be identified, and compared with the opportunity costs, to determine the true return to each resource.

REFERENCES AND RECOMMENDED READINGS

Allen, P. G., L. W. Botsford, A. M. Schur, and W. E. Johnston. 1984. Bioeconomics of Aquaculture. New York: Elsevier Science Publishing Company.

Avault, J. W., Jr. 1985. The alligator story. Aquaculture Mag. 11:41–44.

Bird, K. T. and P. H. Benson. 1987. Seaweed Cultivation for Renewable Resources. New York: Elsevier Science Publishing Company.

Boyd, C. E. 1982. Water Quality Management for Pond Fish Culture. New York: Elsevier Science Publishing Company.

Brown, E. E. 1977. World Fish Farming: Cultivation and Economics. Westport, CN: AVI Publishing Company, Inc.

Cauvin, D. M. and P. C. Thompson. 1977. Rainbow trout farming: an economic perspective. Fisheries and Environment Canada, Fisheries and Marine Service Industry Rep. No. 93.

Davis, J. T. 1986. Baitfish. In Culture of Nonsalmonid Freshwater Fishes, ed. Robert R. Stickney, pp. 149–158. Boca Raton, FL: CRC Press, Inc.

DeVoretz, D. 1982. The uses and abuses of econometric models in the seafood industry: A case study of the B.C. roe-herring industry. Paper presented at the International Seafood Trade Conference, September 8–11, Anchorage, AK.

Engle, C. R. 1987. Optimal product mix for integrated livestock–fish culture systems on limited resource farms. J. World Aquac. Soc. 18:137–147.

Gates, J. M., C. R. MacDonald, and B. J. Pollard. 1980. Salmon culture in water reuse systems: An economic analysis. University of Rhode Island Marine Tech. Rep. 78.

Hambry, J. 1980. The importance of feeding, growth, and metabolism in a consideration of the economics of warm water fish culture using waste heat. In Proceedings of the World Symposium on Aquaculture in Heated Effluents and Recirculation Systems, ed. Klaus Tiews, pp. 601–617. Berlin: H. Heenemann.

Huner, J. V. 1985. An update on crawfish aquaculture. Aquaculture Mag. 11:33,36–40.

Israel, D. C. 1987. Comparative economic analysis of prawn hatcheries. SEAFDEC Asian Aquaculture 9:3–4,12.

Israel, D. C. 1987. Economic feasibility analysis of aquaculture projects: A review. SEAFDEC Asian Aquaculture 9:6–9.

Kofuku, T. and H. Ikenoue, eds. 1983. Modern Methods of Aquaculture in Japan. New York: Elsevier Science Publishing Company.

Korringa, P. 1976. Farming Marine Fishes and Shrimps. New York: Elsevier Science Publishing Company.

Korringa, P. 1976. Farming Marine Organisms Low in the Food Chain. New York: Elsevier Science Publishing Company.

Korringa, P. 1976. Farming the Cupped Oysters of the Genus Crassostrea. New York: Elsevier Science Publishing Company.

Korringa, P. 1976. Farming the Flat Oysters of the Genus Ostrea. New York: Elsevier Science Publishing Company.

Lutz, R. A., ed. 1980. Mussel Culture and Harvest: A North American Perspective. New York: Elsevier Science Publishing Company.

McVey, J. P. and J. R. Moore. 1983. Crustacean Aquaculture. Boca Raton, FL: CRC Press, Inc.

Morse, D. E., K. K. Chew, and R. Mann, eds. 1984. Recent Innovations in Cultivation of Pacific Molluscs. New York: Elsevier Science Publishing Company.

New, M. B., ed. 1982. Giant Prawn Farming. New York: Elsevier Science Publishing Company.

Parker, P. 1988. Salmon farming on the sunshine coast. Seafood Business 7:76–84.

Pitcher, T. J. 1982. Fisheries Ecology. Westport, CT: AVI Publishing Company, Inc.

Shang, Y. C. 1974. Economic feasibility of fresh water prawn farming in Hawaii. Sea Grant Advisory Report, UNIHI-SEAGRANT-AR-74-05.

Shang, Y. C. 1981. Aquaculture Economics. Boulder, CO: Westview Press.

Shang, Y. C. 1981. A comparison of rearing costs and returns of selected herbivorous, omnivorous, and carnivorous aquatic species. Marine Fisheries Rev. 43:23–24.

Shang, Y. C. 1984. Economic aspects of aquafarm construction and maintenance. In Inland Aquaculture Engineering. Lectures Presented at the ADCP Inter-Regional Training Course in Inland Aquaculture Engineering, Budapest, June 6 to September 3, 1983, pp. 539–550. Rome: Food and Agriculture Organization/United Nations Development Programs.

Shigekawa, K. J. and S. H. Logan. 1986. Economic analysis of commercial hatchery production of sturgeon. Aquaculture 51:299–312.

Sindermann, J. C., ed. 1977. Disease Diagnosis and Control in North-American Marine Aquaculture. New York: Elsevier Science Publishing Company.

Vondruska, J. 1976. Aquacultural Economics Bibliography. National Oceanic and Atmospheric Administration Tech. Rep. NMFS SSRF-703.

Waas, B. P., K. Strawn, M. Johns, and W. Griffin. 1983. The commercial production of mudminnows (Fundulus grandis) for live bait. Texas J. Sci. 35:51–60.

Chapter 6

Water and Health Management

WATER MANAGEMENT

Quality

The quality of water is the sum of its chemical and physical characteristics, including dissolved minerals and gases and suspended particulate matter. Quality varies with location, primarily due to geology. Rainwater carries airborne pollutants that may increase the metal or mineral content, and water is altered during surface flow or underground percolation by exchange of ions and compounds, and by the addition of dissolved and particulate material from vegetation and soil. The types and amounts of dissolved or suspended organic or inorganic particles are determined by interactions of water with rock, soil, and the total local environment.

Freshwater supplies are often divided into the categories of hard water and soft water. Hard water has high concentrations of dissolved minerals or ions that often allow fish to recover more rapidly from stresses associated with culture and to increase their tolerance to some pollutants, irritants, and toxicants. Alkalinity is a measure of buffering capacity. It may be more important than hardness, especially in water supplies that are composed of several small supplies or that have a periodic influx of water that significantly differs in pH. The manager should establish that the ion and metal concentrations of the water supply fall within the EPA guideline limits (see Table 1-2).

Important physical characteristics of water include temperature, dissolved gas content, and suspended solids. Temperature has an extreme effect on production, and each species is most efficiently reared in an optimum temperature range (see Table 1-3). Temperature also affects the concentration of dissolved gases, especially oxygen. Equilibrium concentrations decrease as temperatures rise, resulting in less oxygen available for respiration. Every aerobic organism has a required minimum oxygen concentration, below which it dies. When water temperature rises, excess gases cannot escape instantly, and the water becomes supersaturated. Suspended solids can cause gill irritation (Anderson et al. 1984), provide substrates for pathogens, and increase turbidity, resulting in less efficient feeding and growth.

Dissolved materials constitute the ionic or chemical portion of water quality. The pH, directly related to changes in the concentrations of carbon dioxide, ammonia, and carbonate, should generally be between 6 and 9. Alkalinity is a measure of the capacity to accept free hydrogen ions, whereas hardness reflects the concentration of calcium and magnesium ions. Both are measured in terms of a calcium carbonate ($CaCO_3$) equivalent concentration. Dissolved heavy metals, such as copper, lead, and zinc, can cause mortality in relatively' low concentrations (a few parts per million), especially in soft water.

Aeration, Oxygen Injection, Atmospheric Exchange

Aeration is used to increase dissolved oxygen (DO) and to reduce gas supersaturation and concentrations of heavy metals. Oxygen levels decrease in ponds due to biological demands of nighttime plant respiration, fish respiration, and uptake by benthic or other organisms. Mechanical aeration usually requires energy to operate an aerator, and therefore should be limited to times of need. In locations with sufficient relief or gravity, the fall of water can be used for reliable, cost-free aeration and reaeration. Water is aerated by splashing or mixing; with the objective of maximizing the water-to-atmosphere interface or surface area. Oxygen, which makes up about 20% of the atmospheric gases, is transferred into water by a "driving force" equivalent to the concentration gradient, or the difference between the actual DO concentration and the DO saturation concentration. As the DO level approaches saturation, the driving force is reduced, thereby reducing transfer efficiency and increasing the energy cost of aeration.

Oxygen injection systems use 95–100% oxygen instead of air, which greatly increases the DO saturation concentration and thus increases the driving force of transfer. In a pure oxygen atmosphere, the saturation concentration of nitrogen approaches zero. The use of an oxygen atmosphere under pressure allows oxygen to be concentrated in water at extremely high levels while it removes much of the nitrogen. Thus, only a small percentage of the water need be oxygenated and mixed with the rest, leaving the entire supply fully oxygenated and partly stripped of dissolved nitrogen gas.

Aeration and oxygen injection use atmospheric exchange to raise the levels of oxygen in rearing waters. Aeration uses the oxygen in air, while oxygen injection systems create an artificial, pure oxygen atmosphere, producing extreme changes in kinetics and equilibrium concentrations, and resulting in beneficial changes in dissolved gas concentrations when the water is returned to normal atmospheric conditions.

Metabolites

Metabolites are the waste products of digestion and catabolism. Fish release carbon dioxide and ammonia into the water. For instance, in salmonids roughly 0.28 pound of carbon dioxide and 0.03 pound of ammonia are produced for every pound of food consumed. Carbon dioxide released into the water lowers the pH, depending on the buffering capacity. Depending on the DO, high levels of carbon dioxide, 10 mg/L or more, can cause cessation of feeding and eventual mortality. Concentrations in ponds usually fluctuate daily because of plant respiration and are highest in the morning.

Ammonia, also released through the fish's gills, is a by-product of protein metabolism (deamination). There is an ionic form (NH_4^+), and the highly toxic un-ionized form (NH_3). The un-ionized portion of ammonia increases about 10-fold for each unit increase in pH; the effects of a given total ammonia concentration thus increase dramatically, for instance, as the pH increases from 7.0 to 8.0. High ammonia levels result in damage to gills and other tissues, loss of growth, increased susceptibility to disease, and eventual death.

Buffering and Toxicity Mediators

Waters of increasing alkalinity and hardness show an increased capacity to buffer (reduce the effects of) pH and ionic changes. Calcium carbonate chemical reactions use up influxes of acidic ions, and sufficient concentrations of calcium carbonate can prevent major changes in pH. Since carbon dioxide and ammonia can be harmful to fish, the manager should be aware of measures to reduce their effects. Both can be mediated by high levels (near or above saturation) of oxygen and by salt. During times of acute metabolite stress, increases in oxygen, and in freshwater systems increases in salinity up to 10%, reduce damage. Also, since much of the metabolite load is directly related to digestion, feeding should be reduced or halted. Any steps that can be taken to reduce production intensification or temperature (slowly), and to increase flow (and cover), will tend to reduce stress.

Complementary Use

Aquacultural and agricultural systems can be integrated to produce ecologically sound, economical methods of production. Integration is accomplished, for instance, in subsistence culture systems by using culture ponds to catch runoff of waste materials from land animal production. The wastes fertilize the pond and support plant and fish production. Nutrient-rich pond water is then used for garden irrigation. Polyculture of plants and animals (rice and crawfish, for instance) is possible in the same pond.

Effluent: Environmental Impacts and Treatment

The impact of effluent from low-intensity culture is relatively small; however, in all systems there is potential for pathogens to be released into the environment and to threaten wild species. Effluent can be sterilized with chlorine or ozone. In highly intensive systems and in facilities that use pelleted diets, unused food and feces are of concern, especially with regard to phosphorus content. Inasmuch as phosphorus is normally the first limiting factor to plant growth in natural freshwater systems, large influxes of phosphorus can, depending on the latitude, cause the rapid eutrophication of natural waters.

Facilities should be designed so that wastes can be concentrated. Most of the solid nutrient materials can be removed from effluents, and it is commonly recommended, or required, that solids be removed to settling lagoons before the effluent is released into natural ecosystems. The solids should be treated as a sewage product. If dissolved nutrients must be removed, biological treatment is usually required, giving rise to the potential for integrated and hydroponic systems. It is also recommended that chemicals used in the culture system be applied to a discrete flow of water that can then be diverted for treatment. Local environmental regulatory codes should be addressed during the predesign period so that the facility's design can incorporate appropriate features.

Managing Metabolites

Observation of fish behavior and routine monitoring of water quality can be used to head off problems of water quality and metabolite build up. Fish metabolite production can be controlled to some extent by the manipulation of culture practices. The amount of metabolites produced is proportional to the amount of food fed; therefore feeding procedures should be managed to ensure the efficient use of food, thereby reducing ammonia production, oxygen demand, and effects on growth. Ammonia concentration often peaks a few hours after fish are fed. Oxygen demand is usually highest near dawn in ponds and shortly after feeding in highly intensive culture systems. Ammonia peaks and oxygen troughs can be reduced for some fish by the use of demand feeders, which tend to spread the daily feeding periods. The methods and techniques that create and reduce metabolite problems can be anticipated, but they are also learned by experience; thus record keeping can be invaluable to the improvement of the culture system procedures.

Site Selection

The first considerations in site selection are the size and water requirement of the proposed facility. Only after suitable water supplies (quantity and quality) have been located should other factors be considered in detail. Ideally, the site should be near the market. It should have suitable soil; a gently sloping terrain for drainage; and sufficient area for facilities, including buildings, access roads, parking, water treatment, and some space (or option for space) for expansion or unanticipated needs. The use of gravity for water supply and aeration saves money and reduces risk. The layout should be compact but allow easy access to all areas—especially to the location of life support systems. Pond shapes should use the area efficiently and should enhance access and harvesting. Soil type is especially important in earthen pond construction.

Marine finfish culture operations are placed in protected areas, usually along variegated coastlines or intercoastal areas. Site selection for marine shrimp requires tidal flats, preferably with a natural source of larvae. Detailed knowledge of local hydrology (water resources on land, including rainfall) and hydrography data on the tides, currents, and water temperatures are crucial.

HEALTH MANAGEMENT

Health management should be an ongoing process, not a series of reactions to diseases. The degree of management required to control the introduction and development of diseases varies with the magnitude of production, with the intensity of production, with the design of the facility, and with risks due to known (and unknown) hazards and to the nature of life support.

With planning, disease problems can often be avoided by reducing stress and preventing contact between disease agents and the cultured organisms. There are four steps for reducing the risk of disease:

1. *Recognition*. Identify the types of disease-causing organisms normally found in the geographic area, or those likely to be imported, and recognize

the locations in the production system where these pathogens are likely to occur.
2. *Planning.* Plan to eliminate pathogen entry by killing the disease organisms or by interrupting their life cycle. Plan and initiate means of control, such as disinfection measures within the facility.
3. *Design.* Design the facility to reduce the risk of pathogen exposure. The ideal, though possibly unrealistic, goal is to eliminate exposure to pathogens. Also, design the facility to reduce, through water quality management, the environmental factors that lead to disease.
4. *Management.* Culture techniques, such as handling, transporting, and feeding, should be sanitary and conducted in a manner that minimizes the effects of stressors.

Recognition

Recognition is having and applying the capability to identify the signs of pathogens that may occur or have previously caused disease in the facility. Information regarding the types and incidence of fish disease-causing organisms may be available through local sources, publications, and government agencies, and should be solicited and compiled during licensing of the operation. Ideally, the manager or staff person will conduct routine diagnostic investigations. At a minimum, the manager must understand some basic concepts—for instance, that organisms such as the bacterium *Aeromonas salmonicida* (which causes furunculosis in fish) or parasites such as *Ichthyopthirius* or *Costia* are ubiquitous (naturally found everywhere). Other pathogens are restricted geographically but are continually spread by the transport of cultured products. In addition to an appreciation of the concepts, the manager needs a system of monitoring for diseases. If that capability is not available in-house, the manager should investigate university, government, and privately contracted support.

Environmental stressors and toxicants within the water system should also be identified, monitored, and controlled. Water quality is critically important, and any new or planned water supply should be tested with a few of the fish or other organisms to be cultured (as a bioassay) before full-scale use is begun. If the bioassay organisms survive and grow, the supply is deemed safe for culture. A safe water supply is of high quality (such as that for fresh water listed in Table 1-2), has a reliable flow, has no wild fish populations or disease-carrying organisms, and is accessible for as near-total control as possible.

Planning: Management of the Water Supply

Much of water supply management deals with the avoidance of problems, and invaluable steps can be taken before hatchery construction begins. Determination of potential or existing hazards at this time may reduce repeated losses of products and labor (time). Identification of hazards varies, depending on the water supply and how it is used. Before production begins, determine water quantity, quality, and temperature regimes, preferably over an annual cycle, to ensure that the water supply will be ample and suitable.

Pathogenic organisms may be in the water, but usually they live in fish or other life forms, so that removal of wild fish and other organisms in the water source is required. Springs or open water sources should be covered or filled with rubble and freed of organisms to avoid the transfer of pathogens by reduction or elimination of their reservoirs. For example, mud bottom intake areas filled with concrete eliminate tubifex worms that are hosts for spores of whirling disease organisms [*Myxobolus (Myxosoma) cerebralis*]. This technique is used to "manage around" and eventually eliminate the disease from the water source. Identification of potential pathogens and proper design of water intake structures can reduce the introduction of disease-causing organisms. Water supplies can be disinfected or sterilized with ultraviolet light irradiation or ozone treatment. These systems are usually too costly to use, but they do provide an alternative.

Gas supersaturation, a stressor commonly found in groundwater or in pumped or heated water supplies, can be reduced or eliminated in a variety of ways. The supersaturated gases such as nitrogen, oxygen, or carbon dioxide equilibrate, or come out of solution, inside the organism and result in the formation of bubbles, a condition similar to "the bends" in deep-sea divers. Gas bubbles are formed near the epidermis and in the circulatory system; acute conditions can be lethal. Supersaturation occurs most often in systems in which water has little retention time, as in intensive culture systems, rather than in extensive systems. Supersaturation can be reduced by passing the water supply through columns packed with plastic structures to break up the flow, in combination with vacuum degassing, oxygen injection, aeration, and retention. The object of each treatment is to remove excess gas, usually nitrogen, while keeping oxygen at an acceptable level for culture. Vacuum systems reduce the oxygen concentration, since they indiscriminately reduce all gases. Injection of pure oxygen results in the dissolved oxygen concentration increasing above ambient saturation while the nitrogen level rapidly decreases well below saturation. Reduction of total gas pressure (the sum of the partial pressures of all the gases) to or below 102% is recommended for fish culture—although gas bubbles are formed in some larvae, at 102%.

Water quality should also be monitored, especially for levels of oxygen and ammonia. Oxygen is usually the first limiting factor to production; ammonia, a metabolite excreted across the gills in fish, is usually the second. Ammonia can reach toxic concentrations in highly intensive culture systems, but toxicity decreases as oxygen concentration increases.

Design

System design should be site specific and geared to production needs. Simplicity reduces operational difficulties; avoiding the use of pumps and power-dependent life support devices significantly reduces the risk of losses caused by systems failure. Design goals should incorporate efficient operation in terms of both labor and water use, as well as a gravity water flow default system or power backup system for life support. Design goals should also include minimization of stresses, prevention of disease, and the capability for isolation of diseased organisms if they occur.

A facility design should include ample work space for conducting production chores easily. Ample storage areas are important, since cluttered or obstructed walkways reduce labor efficiency and can add to the handling stresses on fish. Buildings should be designed with contingencies for change, such as additions.

Management

Health management involves three ingredients: (1) management of the water supply to reduce disease exposure and stress, (2) management of procedures to minimize stresses related to culture activity, and (3) management of personnel to ensure that culture tasks are performed efficiently and that inherent stressor effects are minimized. Management of procedures and personnel will very likely involve training.

Culturists and managers should have a good understanding of the role that stressors play on cultured organisms—especially the relation between stress and increased disease susceptibility. Stress is a condition of the cultured organism, caused by the environment or by a pathogen, that reduces growth or increases disease susceptibility or both. Stressors in fish culture include those of operating procedures, such as normal activity (human) and handling, crowding, transporting, and treating (for disease), as well as environmental factors such as changing temperatures, low oxygen levels, bright light, high ammonia concentration, high gas pressure, contaminants, and others. There are numerous stress-mediated diseases, or those brought on by stress, that lower the organism's defenses (increasing the susceptibility to pathogen invasion). Examples include parasite infections (*Costia, Trichodina,* or *Hexamita*); bacterial diseases (furunculosis, bacterial gill disease, columnaris); and viral diseases (Infectious Hematopoietic Necrosis, channel catfish virus). In each of these infestations, an environmental stressor is a predisposing factor in the infection. Management must strive to reduce the stress of normal culture technique by closely controlling water quality (or water supply characteristics) and fish culture techniques.

In human medicine and personal hygiene, we recognize the overwhelming importance of a healthy environment and cleanliness or sanitation. Likewise, good fish culture management techniques reduce rearing problems. Fish and other cultured aquatic organisms respire in water that contains the excreted wastes and disease agents sloughed by the fish upstream, and many stressors result directly from cultural practices.

Basic hatchery sanitation measures, i.e., disinfection of nets, boots, and other equipment between tanks; are important aspects of disease control too often overlooked. Culturists can reduce risk and increase production quality by carefully monitoring the culture conditions. The weakest link in the management scheme is often the lack of properly trained personnel. Procedures must be established for correctly handling all culture activities, and management must insist on strict adherence to policy. Culturists should be trained in disease recognition, especially for those diseases historically found on the site. Early detection of disease can significantly reduce losses. Good record keeping will result in a historical account of any inherent rearing problems that may exist. Seasonal disease outbreaks can be reduced or eliminated if stressful culture practices are anticipated and avoided before the seasonal disease challenge occurs.

REFERENCES AND RECOMMENDED READINGS

Alabaster, J. S. and R. Lloyd. 1980. Ammonia. *In* Water Quality Criteria for Fresh-water Fish, ed. J. S. Alabaster and R. Lloyd, pp. 85–102. London: Butterworths.

American Public Health Association, American Water Works Association, and Water Pollution Control Federation. 1985. Standard Methods for the Examination of Water and Wastewater, 16th ed. Washington, DC: American Public Health Association.

Anderson, D. P., O. W. Dixon, and F. W. Van Ginkel. 1984. Suppression of bath immunization in rainbow trout by contaminant bath pretreatments. *In* Chemical Regulation of Immunity in Veterinary Medicine, ed. M. Kende, J. Gainer, and M. Chirigos, pp. 289–293. New York: Alan R. Liss, Inc.

Beveridge, M. C. M. 1987. Cage Aquaculture. Farnham, Surrey, England: Fishing News Books Ltd.

Bouck, G. R. 1982. Gasometer: An inexpensive device for continuous monitoring of dissolved gases and supersaturation. Trans. Am. Fish. Soc. 111:505–516.

Bouck, G. R. and R. E. King. 1983. Tolerance to gas supersaturation in fresh water and sea water by steelhead trout, *Salmo gairdneri* Richardson. J. Fish Biol. 23:293–300.

Boyd, C. E. 1979. Water Quality in Warmwater Fish Ponds. Auburn, AL: Auburn University Agricultural Experiment Station.

Brett, J. R. and C. A. Zalla. 1975. Daily pattern of nitrogen excretion and oxygen consumption of sockeye salmon (*Oncorhynchus nerka*) under controlled conditions. J. Fish. Res. Board Can. 32:2479–2486.

Brockway, D. R. 1950. Metabolic products and their effect. Prog. Fish-Cult. 12:127–129.

Brown, E. E. and J. B. Gratzek. 1980. Fish Farming Handbook. Westport, CT: AVI Publishing Company, Inc.

Bullock, G. L. and K. Wolf. 1986. Infectious disease of cultured fishes. Fish and Wildlife Leaflet 6. Washington, DC: U.S. Fish and Wildlife Service.

Burrows, R. E. 1964. Effects of accumulated excretory products on hatchery reared salmonids. U.S. Fish and Wildlife Service Research Report 66. Washington, DC: U.S. Fish and Wildlife Service.

Buss, K. 1979. The Fundamentals of Fish Culture: An Outline for Classroom Study. Mansfield, PA: Mansfield State College Press.

Canadian Council of Resource and Environment Ministers. 1987. Canadian Water Quality Guidelines. Ottawa, Ontario Environment Canada.

Colt, J. E. 1983. The computation and reporting of dissolved gas levels. Water Res. 17:841–849.
Colt, J. E. 1984. Computation of Dissolved Gas Concentrations in Water as Functions of Temperature, Salinity, and Pressure. Special Publication No. 14. Bethesda, MD: American Fisheries Society.
Colt, J. E. and D. A. Armstrong. 1981. Nitrogen toxicity to crustaceans, fish and molluscs. In Bio-engineering Symposium for Fish Culture, ed. L. J. Allen and E. C. Kinney, pp. 34–47. American Fisheries Society Fish Culture. Sect. Pub. 1.
D'Aoust, P. -Y. and H. W. Ferguson. 1984. The pathology of chronic ammonia toxicity in rainbow trout, *Salmo gairdneri* Richardson. J. Fish Dis. 7:199–205.
Dwyer, W. P. and R. G. Piper. 1984. Atlantic salmon growth efficiency as affected by temperature. Bozeman Information Leaflet 30. Washington, DC: U.S. Dept. of the Interior, Fish and Wildlife Service.
Ebel, W. J., K. T. Beiningen, G. R. Bouck, W. R. Penrose, and D. E. Weitkamp. 1979. Gases, total dissolved. In A Review of the EPA Red Book: Quality Criteria for Water, ed. R. V. Thurston, R. C. Russo, C. M. Fetterolf, T. A. Edsall, and Y. M. Barber, pp. 113–118. Bethesda, MD: American Fisheries Society.
Emerson, K. R., R. C. Russo, R. E. Lund, and R. V. Thurston. 1975. Aqueous ammonia equilibrium calculations: Effect of pH and temperature. J. Fish. Res. Board Can. 32:2379–2383.
European Inland Fisheries Advisory Committee. 1980. Water quality criteria for European fresh-water fish. Report on combined effects on fresh-water fish and other aquatic life on mixtures of toxicants in water. EIFAC Technical Paper No. 37. Rome: Food and Agriculture Organization of the United Nations.
Gunther, D. C., D. E. Brune, and G. A. E. Gall. 1981. Ammonia production and removal in a trout rearing facility. Trans. Am. Soc. Agric. Eng. 24:1376–1380, 1385.
Jensen, J. O. T., J. Schnute, and D. F. Alderdice. 1986. Assessing juvenile salmonid response to gas supersaturation using a general multivariate dose-response model. Can. J. Fish. Aquat. Sci. 43:1694–1709.
Jobling, M. 1981. Some effects of temperature, feeding, and body weight on nitrogenous excretion in young plaice, *Pleuronectes platessa* L. J. Fish Biol. 18:87–96.
Klontz, G. W. 1979. Fish Health Management, Vols. 1 and 2. Boise, ID: University of Idaho.
Liao, P. B. 1974. Ammonia production rate and its application to fish culture system planning and design. Paper read at the 25th Northwest Fish Culture Conference, December 4–6, 1974, pp. 107–119. Seattle, WA.
Marking, L. L. 1987. Gas supersaturation in fisheries: Causes, concerns, and cures. Fish and Wildlife Leaflet 9. Washington, DC: U.S. Fish and Wildlife Service.
Meade, J. W. 1985. Allowable ammonia for fish culture. Prog. Fish-Cult. 47:135–145.
Meyer, F. P., J. W. Warren, and T. G. Carey, eds. 1983. A Guide to Integrated Fish Health Management in the Great Lakes Basin. Special Pub. 83–2. Ann Arbor, MI: Great Lakes Fishery Commission.
Meyers, T. R. and J. D. Hendricks. 1982. A summary of tissue lesions in aquatic animals induced by controlled exposures to environmental contaminants, chemotherapeutic agents, and potential carcinogens. Mar. Fish. Rev. 44:1–17.
Nebeker, A. V., G. R. Bouck, and D. G. Stevens. 1976. Oxygen and carbon dioxide and oxygen–nitrogen ratios as factors affecting fish in air-supersaturated water. Trans. Am. Fish. Soc. 105:425–429.
Nebeker, A. V., A. K. Hauck, and F. D. Baker. 1979. Temperature and oxygen–nitrogen gas ratios affect fish survival in air-supersaturated water. Water Res. 13:299–303.
Paulson, L. J. 1980. Models of ammonia excretion for brook trout (*Salvelinus fontinalis*) and rainbow trout (*Salmo gairdneri*). Can. J. Fish. Aquat. Sci. 37:1421–1425.

Peterson, H. 1971. Smolt rearing methods, equipment and techniques used successfully in Sweden. Atlantic Salmon Foundation Special Publication Series 2:32–62.

Piper, R. G., I. B. McElwain, L. E. Orme, J. P. McCraren, L. G. Fowler, and J. R. Leonard. 1982. Fish Hatchery Management. Washington, DC: U.S. Fish and Wildlife Service.

Plumb, J. A. 1981. Relationship of water quality and infectious diseases in cultured channel catfish. In Fish, Pathogens and Environment in European Polyculture, pp. 290–303. Szarvas, Hungary.

Poxton, M. G. and S. B. Allouse. 1982. Water quality criteria for marine fisheries. Aquacult. Eng. 1:153–191.

Pullin, R. S. V. and Z. H. Shehadeh. 1980. Integrated Agriculture-Aquaculture Farming Systems. Manila, Philippines: International Center for Living Aquatic Resources Management.

Rucker, R. R. 1976. Gas bubble disease of salmonids: Variation In oxygen–nitrogen ratio with constant total gas pressure. In Gas Bubble Disease, ed. D. H. Fickeisen and M. J. Schneider, pp. 85–88. Oak Ridge, TN: Conf-741033, Technical Information Center, Energy Research and Development Administration.

Smith, C. E. and R. G. Piper. 1975. Lesions associated with chronic exposure to ammonia. In The Pathology of Fishes, ed. W. E. Ribelin and G. Migaki, pp. 497–514. Madison, WI: University of Wisconsin Press.

Soderberg, R. W. 1982. Aeration of water supplies for fish culture in flowing water. Prog. Fish-Cult. 44:89–93.

Tomasso, J. R., C. A. Goudie, B. A. Simco, and K. B. Davis. 1980. Effects of environmental pH and calcium on ammonia toxicity in channel catfish. Trans. Am. Fish. Soc. 109:229–234.

Van Banning, P. 1987. Second International Colloquium on Pathology in Marine Aquaculture. Aquaculture 67(1/2).

Visscher, L. and W. Godby. 1987. Oxygen Supplementation: A New Technology in Fish Culture. Information Bull. No. 1. Denver, CO: U.S. Fish and Wildlife Service.

Wedemeyer, G. 1970. The role of stress in the disease resistance of fishes. In A Symposium on Diseases of Fishes and Shellfishes, ed. S. F. Snieszko, pp. 30–35. American Fisheries Society Spec. Pub. No. 5.

Wedemeyer, G. A. and W. T. Yasutake. 1977. Clinical methods for the assessment of the effects of environmental stress on fish health. Techn. Paper No. 89. Washington, DC: U.S. Fish and Wildlife Service.

Westers, H. 1983. Considerations in hatchery design for the prevention of diseases. In A Guide to Integrated Fish Health Management in the Great Lakes Basin, ed. F. P. Meyer, J. W. Warren, and T. G. Carey, pp. 29–35. Ann Arbor, MI: Great Lakes Fishery Commission.

Willingham, W. T., J. E. Colt, J. A. Fava, B. A. Hillaby, C. L. Ho, M. Katx, R. C. Russo, D. L. Swanson, and R. V. Thurston. 1979. Ammonia. In A Review of the EPA Red Book: Quality Criteria for Water, ed. R. V. Thurston et al., pp. 6–18. Bethesda, MD: Water Quality Section American Fisheries Society.

Wolf, K. 1969. Blue-sac Disease of Fish. Fish Disease Leaflet 15. Washington, DC: U. S. Dept. of the Interior.

Wood, J. W. 1968. Diseases of Pacific salmon, Their Prevention and Treatment. Olympia, WA: Washington Department of Fisheries, Hatchery Division.

Wright, P. B. and W. E. McLean. 1985. The effects of aeration on the rearing of summer chinook fry (Oncorhynchus tshawytscha) at the Puntledge Hatchery. Can. Tech. Rep. Fish. Aquat. Sci. 1390.

Chapter 7

Ethics

FOCUS:	The manager is responsible for developing and following ethical approaches to business and environmental resources and, in the public sector, maintaining the integrity of publicly funded projects.
HIGHLIGHTS:	• Ethical standards in the workplace • Utilitarianism and the rights view • Sentience versus needs • Fish lack a right to life • Ethics and effectiveness
ORGANIZATION:	Business and Government Environmental Ethics: From Management of Scarce Natural Resources to Commodity Production for Human Consumption Moral Standing Animal Rights Preserving Rare Species Conclusions Professionalism and Legislated Ethics (for Public Domain Managers)

BUSINESS AND GOVERNMENT

It may have been said that a book entitled *Ethics for the Aggressive Businessman* must be one of the shortest ever printed. In fact, in brevity one might expect it to be comparable to *Great Attributes All In-Laws Possess*. There is, however, a substantial literature on business ethics. Topics include duties to the public sector, allocation decisions, and the conditions, if any, under which it is morally permissible to lie, withhold information, coerce, whistle blow, preferentially hire, or disobey the law.

In the public sector, accepted ethical standards require that managers and employees avoid any action that might result in or create the appearance of:

Using public office for private gain.
Giving preferential treatment to any organization or person.
Impeding government efficiency or economy.
Losing independence or impartiality of action.
Making a government decision outside official channels.
Reducing the confidence of the public in the integrity of the government.

The manager must deal with questions of outside employment, financial interests, gifts, disclosure of information, political activity, and whistle blowing versus the right of privacy. However, virtually all federal and large state or provincial government agencies have regulations and explanatory information regarding ethical principles, and most have counselor and solicitor services. Unquestionably, managers in these systems should, and sometimes must, take advantage of specialists trained to deal with these "gray area" concerns.

ENVIRONMENTAL ETHICS: FROM MANAGEMENT OF SCARCE NATURAL RESOURCES TO COMMODITY PRODUCTION FOR HUMAN CONSUMPTION

A method used by philosophers to resolve moral problems involves applying an ethical theory to those problems. To do so, one of many competing ethical theories (or decision-making procedures) must be adopted. Selecting a theory requires substantial effort. Here we consider two widely divergent and well-known approaches: utilitarianism, a type of consequentialism, and the rights view, a type of non-consequentialism.

The utilitarian principle asserts that an action is right if it is expected to maximize net utility; otherwise, it is wrong. Not all utilitarians agree on the proper method for a cost-benefit analysis, but all agree that the consequences of an action determine its moral worth. Somehow we must aggregate the pleasures and pains expected from each alternative action, and determine which action will produce the greatest good (pleasure) and the least evil (pain). Everyone's pleasures and pains count (or, in the language of contemporary economic theorists and preference utilitarians, everyone's satisfied and frustrated preferences count). Many utilitarians believe that talk of moral rights is "nonsense on stilts." Utilitarians who use such language point out that moral rights do not exist independently of, and ultimately are derived from, the principle of utility; to say that "X has a right to life" is shorthand for "killing X would violate the principle of utility."

Proponents of the rights view contend that moral rights exist independently of utilitarian considerations. If X has a right to life, actions that violate that right cannot be morally justified by claiming that they maximize net utility. Even an action that produces a great deal of happiness and no suffering is morally wrong if it involves a violation of a right. That is the only way, argue rights theorists, to avoid counterintuitive utilitarian conclusions such as "Killing and stealing are all right as long as they maximize utility. "

On the rights view, *reflective intuitions* (or *considered beliefs*) function as the data against which theories are tested. Intuitions are reflective only if we have thought about them impartially, while in a normal frame of mind, and with full information. Reflection on unproblematic cases (e.g., if you had the choice of killing 50 trapped miners to save 2, killing 2 to save 50, or doing nothing and allowing all 52 to die, what would you do?) gives rise to a base of considered beliefs; from that base, general principles are constructed that unify those beliefs by identifying plausible common ground. Then an assumption is made that these general principles can be applied to problematic cases—ones we are not sure about—yielding moral judgments.

A typical utilitarian response is that though some versions of utilitarianism lead to counterintuitive conclusions, other versions (when properly understood) do not. In addition, any moral theory will entail some moral judgments that strike us as mistaken; however, moral theories are designed to correct our intuitions and should not be based on them (Regan 1983). On many issues utility-based and rights-based approaches lead to the same conclusions.

Moral Standing

Do fish have "moral standing"? More generally, what set of characteristics must something possess to be a moral rights holder or to deserve considera- tion in a cost-benefit analysis? Traditionally, being a member of the species *Homo sapiens* was widely considered a necessary and sufficient condition. Anthropocentrism has come under harsh attack, and since 1975 considerable scholarship has been devoted to animal liberation. Though they differ on many points and adhere to a range of irreconcilable moral theories, animal libera- tionists agree that it is arbitrary and indefensible to discount the impact of human actions on nonhumans solely on the ground that those impacted are not members of our species. Species membership, they contend, is a morally irrelevant trait, much like race and sex.

Sentience, or the capacity to be aware of pleasure and pain, has now become a widely held criterion of moral standing. Utilitarian sentientists support the principle of equality: the pleasures and pains of *any* being that are expected to result from a human action must be included in the cost-benefit analysis of that action, and must be given equal weight with the like pleasures and pains of any other affected being. Rights-based sentientists contend that all sentient beings have a moral right to be spared undeserved suffering.

Even if it is reasonable to adopt sentience as a criterion of moral standing, one might argue that it is unlikely that fish can suffer. The only way to support the belief that other species are sentient is through analogical reasoning. One comes to believe that other humans feel pain by observing their behavior (such as moaning, grimacing, or physiological changes) and inferring that, since they are physiologically similar to oneself, and exhibit similar behavior when similarly stimulated (pain), in all probability they can feel pain. When applied to nonhuman animals, this strategy supplies an inductive basis for believing that at least some animals also feel pleasure and pain. However, the conclusion of an analogical argument receives little support from its premises when the two species being compared have little in common, such as humans and fish. This does not mean that fish are insentient, but only that, on the basis of analogical reasoning, we have no compelling reason to believe otherwise.

One might argue that fish undergo physiological changes similar to those present in humans experiencing pain; thus, since in humans there is a constant correlation between feeling pain and, say, physiological changes x, y, and z, it is reasonable to conclude that when fish undergo changes x, y, and z, they too are feeling pain. However, the mere presence of changes, such as the ammonia content of muscles or synaptic responses (to burns, injuries, etc.), is not a sufficient condition for sentience. To be sentient, a being must not only be capable of undergoing changes, but must *consciously* experience those changes, and that probably requires a rather highly developed brain. Physiological monitoring may pick up changes x, y, and z, but what do those changes mean to the fish? They may simply indicate that the fish is adapting to its environment. Since there are great differences between the brains of humans and fish, we have, at best, only a weak analogical basis for believing that fish are capable of translating changes x, y, and z into a subjective experience of pain. The belief that fish are insentient should be abandoned, however, if a brain pattern in humans is discovered which always occurs when humans report being in pain, and an identical brain pattern is found to occur in fish. In that case, one should consider it probable that fish are sentient.

Though most sentientists believe that fish are insentient, they do not deny that fish have needs, wants, and interests that can be frustrated by human actions. It makes sense to say that fish "need" or "have an interest in" clean water, that salmon "want" to reach suitable spawning grounds, and that dams, pollution, and other factors can "frustrate" those needs, wants, and interests (henceforth, *desires*). Although the effects of human activity on fish and aquatic ecosystems can be appalling, most sentientists will not count the frustrated desires of those fish as a "cost," because they believe that fish *feel* no pain. Present and future humans and other sentient beings may suffer as a result of what happens to fish and aquatic ecosystems, and sentientists will count that suffering. Various absurdities follow if frustrated or satisfied desires, rather than the *awareness* of frustrated or satisfied desires, are taken as a criterion of moral standing or a basis for moral rights. If we claim that a fish's desire to spawn, feed, or avoid excessive dissolved nitrogen gas has moral standing, though it does not feel pain when these desires are frustrated, by parity of reasoning we must conclude that a plant's desire for water has moral standing—which means that depriving it of water constitutes a rights violation or a cost. If we allow unconscious desires to have moral standing, and thus include the frustration and satisfaction of such desires in cost-benefit calculations, we are logically committed to the view that plants have moral standing, as do disease-causing microbes or possibly even self-regulating, goal-directed machines such as missiles, which "want" to hit their target (Frey 1986). Counterintuitive implications such as these lead most of us to abandon the idea that frustrated and satisfied unconscious desires have moral standing. This is not to deny that they might matter to us. Many of us deeply respect the complexity and beauty of such desires in biological organisms, and even go to great lengths to remove impediments to their fulfillment.

Animal Rights

Is it possible that fish have inherent value or a right to life, and therefore that killing them would be morally permissible only if one had reasons strong enough to override that moral right? The two most influential and widely discussed theories, delineating the features a being must possess in order to have a right to life, clearly indicate that fish do not have that right. Perception, belief, memory, desire, self-consciousness, and a sense of the future, including one's own, are the attributes required in both the rights-based (Regan 1983) and utility-based (Singer 1986) approaches to establish a right to life or (as utilitarians prefer to say) are necessary if one is to be covered by our prohibition against killing.

One of the few ways of arriving at the conclusion that fish have a right to life is by supporting the claim, as did Albert Schweitzer, that all living organisms have a moral right to life. Overlooking the significant difficulties involved in trying to show that merely being alive is a necessary and sufficient condition for a right to life, an enormous practical problem of this view is the need to construct a decision-making procedure to adjudicate the innumerable conflicts that occur. One must decide if all living things have an equal right to life (equal inherent value); if so, it is just as wrong, side effects excluded, to kill a fish or a smallpox virus as a human being. To avoid such implications, a hierarchical system can be adopted, with the right to life of some beings overriding that of others; but how are such rights to be nonarbitrarily ranked, and can a collection of lower-ranking rights override one with a higher rank? Many other questions arise, and most philosophers conclude that biotic egalitarianism is based on mysticism rather than logic. However, a small group of environmental ethicists contend that such a position is coherent (Rodman 1977, Goodpaster 1978, Callicott 1986, Taylor 1986).

Preserving Rare Species

Is it morally wrong to knowingly exterminate or endanger a native species or subspecies (say, by introducing hatchery-reared fish or a non-native species into their habitat)? The claim that such an action can be morally wrong is difficult to justify in terms of the rights or inherent value of a species (as opposed to individual members of a species). No plausible criteria for attributing rights to individuals (such as rationality, sentience, or autonomy) apply to species. Nearly all rights-based and utility-based moral theorists agree that it is best to approach the preservation issue by focusing on individuals. For example, do individual Alvord cutthroat trout have inherent value, or instrumental value, and under what conditions, if any, is it morally justified to kill a being with inherent value?

As noted, the rights view and similar positions hold that fish lack a right to life. Thus, they are of no help in deciding the morality of attempting to preserve members of endangered species. Utilitarianism, however, does supply a decision-making procedure capable of leading to the conclusion that exterminating or endangering a native species is morally wrong, namely, when doing so fails to maximize expected net utility.

There are significant differences among utilitarians on how to define *utility*, or, in other words, on how to conduct a cost-benefit analysis properly. Most argue that it is unjustified to give weight in the calculations only to the pleasures and pains of humans; some claim that not all pleasures and pains should count—after all, some are clearly irrational (as the products of false beliefs, psychological insecurities, indoctrination, drugged states, etc.), and justification for their inclusions in a cost-benefit analysis is unclear. Also, the weight to be given to interests of future generations is hotly disputed. Hierarchical claims and interests of our progeny are only two of a host of related complex issues. Rather than address specific issues in detail, it might be more useful to sketch the method by which a typical utilitarian would evaluate a project—say, a proposed dam that would eliminate a race of magnificent steelhead.

According to utilitarianism, the members of a species or a subspecies have no inherent value but do have instrumental value. It makes no sense to argue that it is morally wrong to build the dam by claiming that steelhead have a right to continued existence. Courts may someday give them legal rights (Stone 1974), but there appears to be no rational basis for attributing moral rights to them. Dam construction is morally wrong only if it fails to maximize expected utility, that is, if it is likely to produce less net good (utility) in the long run (the pleasures and pains of all affected sentient beings considered) than the alternative action. Factors that must be considered in a cost-benefit analysis include the aesthetic, economic, ecological, and scientific values of the steelhead. Several questions must be answered, such as: How many jobs will be created and lost, and how will that factor "cash out" in terms of human happiness and suffering? How will the fish loss affect nonhuman sentient beings which depend on them? How will that loss affect humans, not only those near the river but also those far away who have an interest in what happens? And will other fishing or lake recreational opportunities increase, and how will this increase affect local business, hatchery employees and their families, and so on? Aversions to water skiing, for instance, need to be included as a cost of the project. One must be impartial, not giving extra weight to one's own interests; equal interests count equally, no matter whose they are (male or female, black or white, rich or poor, human or nonhuman, mine or yours). Also, one might want to exclude irrational happiness and suffering from the calculations—for instance, the happiness and suffering of someone who vows to commit suicide if the dam is not constructed—and if so, a criterion must be developed. Many other issues arise, including the problem of measuring pleasure and pain. Yet, despite the significant theoretical and practical problems of conducting an acceptable cost-benefit analysis, utilitarians maintain that there is no better method of resolving conflicts of interest.

The utilitarian reasoning does not suggest that we should do all we can to prevent the extermination of any species. Whether we should attempt to save a species, and the amount of effort to expend, depends on the outcome of the cost-benefit analysis. It is quite possible that taking the necessary steps to save a species, or a unique population, is not worth the effort—and that possibility is disconcerting to preservationists.

Preservationists make a case for the inherent value of the fish or ecosystem. If members of a species have inherent value, it becomes morally difficult to justify an action that reduces their numbers, even if that action would make many people much better off and produce little harm. An increase in net utility does not justify killing a being with a right to life; if it did, it could be morally justified for a person to kill his wealthy, misanthropic aunt (who is about to die anyway) and distribute her money among the needy. A right to life takes precedence over the results of a cost-benefit analysis. If fish were to have a right to life, our relation to them would change dramatically. Precisely how it would change would depend on the strength of that right. If it could be easily overridden or outweighed by other rights or even by utilitarian considerations, we would continue harvesting fish, building dams, and so on. If, however, every instance of killing a fish would raise a serious moral issue, the issues central to the morality of killing any being with a right to life would come into play. If X has a right to life, does that entail a positive duty to take steps to prevent X from dying, or merely a negative duty to refrain from killing X? Is the failure to save a being with a right to life morally equivalent to killing such a being? Some preservationists, for example, argue that we have a positive duty to preserve most or all species and subspecies, in addition to a negative duty not to exterminate them. If so, the failure to take adequate steps to enhance problem populations is morally impermissible.

Conclusions

There are two central ethical questions of importance to culture systems management: Under what conditions is it morally justified to deliberately cause pain to a being? Under what conditions is it morally justified to deliberately kill a being? If fish are insentient, the first question is of no concern, and if fish lack a right to life, the second question is of no concern. Determining whether fish are sentient is more than just an empirical issue; for even if all of the relevant physiological facts are someday known, one must still construct an analogical argument and decide what degree of probability the premises lend to the conclusion. If one concludes that fish are sentient, one must develop a response to the first question. Determining whether fish have a right to life involves philosophical and empirical issues: We must select criteria of a right to life, and then study fish to see if they satisfy those criteria. One method of generating criteria involves reflecting on paradigmatic cases of rights holders (adult humans) and isolating one or more relevant characteristics they all possess; a characteristic is deemed relevant to X having a right to life if, in the absence of that characteristic, one would conclude that X lacks a right to life. Though useful, this method is not theory neutral, as the characteristics one selects are heavily influenced by one's moral theory. For instance, many Christian ethicists would consider possessing a soul a necessary and sufficient condition for possessing a right to life, whereas most utilitarians would isolate the preference for continued existence (which presupposes a conception of one's own mortality). Either way, as with nearly all moral theories, fish lack a right to life.

PROFESSIONALISM AND LEGISLATED ETHICS (FOR PUBLIC DOMAIN MANAGERS)

Professionalism is variously defined to include those in learned disciplines, those who devote themselves to a chosen field, and those who will do the right thing or that which they said they would do, no questions asked. Further, all professionals are assumed to be standard bearers for ethics and integrity in their fields. There are cynics who tend to think that professional include only medical doctors, attorneys, and those in other fields involving nonmanual work who make a lot of money—but the cynics number no more than 180–190 million in the United States. Among fisheries workers, there seems to be an undertone of anxiety in discussions of professionalism. Might we try, too hard sometimes, to convince ourselves that we are "real" professionals? Could we be among the cynics? To paraphrase Shakespeare, "the fishery manager doth protest too much, me thinks." There is no clear picture or image of professionalism among fish culturists. There is, however, a healthy admiration for hard work and understated accomplishment.

Certain moral opinions may be held by a large number of people in the United States (or at least by their elected representatives) regarding interjurisdictional questions of fish management and questions regarding threatened and endangered fishes. For instance, the U. S. Congress has directed the Department of the Interior, Fish and Wildlife Service, to manage and conserve natural resources. Thus, with some interpretation, managers in public sector conservation agencies have guidelines for moral valuation. These public managers are to prevent (by interest groups to the denial of others, and of posterity) the extinction of native species. Even though such a view may presently not be supported by a logically well-founded argument, it is the view that public managers must own or "buy into" if they are to conduct their appointed duties.

However, development of national policies for conservation, of species or of any resources, is reasonable and not without logic. As colonies assumed statehood (in the United States), each exercised the prerogative and accepted the responsibility of managing its resources. The management of some resources common to two or more states, such as fisheries, became a chronic problem, and the states looked to the federal government for a solution. In 1871 the U. S. Congress established the Office of Commissioner of Fish and Fisheries because of mounting concerns about the widespread decline in the nation's food fish supply. Federal stewardship of national fisheries resources has grown and changed, but three factors continue to be centrally important in shaping federal fishery responsibilities: (1) a seemingly ever-growing awareness of the importance of fishery resources as sources of food and recreation; (2) a greater recognition and understanding of the limits of fishery resources, and of the adverse effects of overfishing and habitat damage; and (3) a need for comprehensive, coordinated management as interests, interactions, and technology become more sophisticated.

U. S. Fish and Wildlife Service managers now have the responsibilities to:

Facilitate the restoration of depleted, nationally significant fishery resources.
Seek and provide for mitigation of fishery resources impaired by federal water-related development.
Assist with management of fishery resources of federal Indian lands.
Maintain a federal leadership role in scientifically based management of national fishery resources.

Much like law enforcement personnel, public managers in fisheries are given a set of legislated values and directed to accomplish certain work based on those values. Their work does not address whether a species or population of fish has a moral right, but rather with how to apply resources to exercise the mandate to protect those fish. Ethical questions for the public manager, then, focus on effectiveness. For instance, if a restoration (mitigation) hatchery is producing a negative (rather than positive) impact on the target population, how can the manager justify its operation? Are resources being applied for maximum effectiveness? And if not, is the manager compromising a public trust?

REFERENCES AND RECOMMENDED READINGS

Arrow, K. J. 1973. Social Choice and Individual Values. New Haven, CT: Yale University Press.

Arrow, K. J. and A. C. Fisher. 1974. Environmental preservation, uncertainty, and irreversibility. Q. J. Econ. 88:313–319.

Baier, K. 1977. Rationality and morality. Erkenntris 11:197–223.

Baumol, W. J. and W. E. Oates. 1975. The Theory of Environmental Policy. Englewood Cliffs, NJ: Prentice-Hall, Inc.

Bell, F. W. 1968. The pope and the price of fish. Am. Econ. Rev. 57:1346–1350.

Bell, F. W. 1972. Technological externalities and common property resources. J. Pol. Econ. 80:148–158.

Benson, G. 1982. Business Ethics in America. Lexington, MA: Lexington Books.

Bockstael, N. E. and K. E. McConnell. 1981. Theory and estimation of the household production function for wildlife recreation. J. Environ. Econ. Manag. 8:199–214.

Boyet, W. E. and G. S. Tolley. 1966. Recreation projects based on demand analysis. J. Farm Econ. 48:4.

Braybrooks, D. 1983. Ethics in the World of Business. Totawa, NJ: Rowman and Allenheld Publishers.

Brookshire, D. S. , L. S. Eubanks and C. F. Sorg. 1986. Existence values and normative economics: Implications for valuing water resources. Water Resour. Res. 22:1509–1518.

Brookshire, D. S. , L. S. Eubanks, and A. Randall. 1983. Estimating option prices and existence values for wildlife resources. Land Econ. 59:1–15.

Brown, B. 1982. Mountain in the Clouds. New York: Simon and Schuster.

Burt, O. R. and D. Brewer. 1971. Estimation of net social benefits from outdoor recreation. Econometrica 39.

Callicott, J. B. 1986. The search for an environmental ethic. In Matters of Life and Death: New Introductory Essays in Moral Philosophy, ed. Tom Regan, pp. 381–424. New York: Random House.

Callicott, J. B. 1986. Animal liberation: A triangular affair. In People, Penguins, and Plastic Trees: Basic Issues in Environmental Ethics, ed. Donald Van De Veer and Christine Pierce, pp. 184–203. Belmont, CA: Wadsworth Publishing Company.

Cesario, F. J. and J. L. Knetch. 1976. A recreation site demand and benefit estimation. Regional Studies 10:97–104.

Clawson, M. 1959. Methods of measuring the demand for the value of outdoor recreation. RFF Reprint 10. Washington, DC: Resources for the Future.

Clawson, M. and J. Knetch. 1966. Economics of outdoor recreation. Baltimore, MD: Johns Hopkins University Press.

Cummings, R. G. , D. S. Brookshire, and W. D. Schulze. 1986. Valuing Environmental Goods: An Assessment of the Contingent Valuation Method. Totawa, NJ: Rowman and Allenheld Publishers.

Daubert, J. T. and R. A. Young. 1979. Recreational demands for maintaining instream flows: A contingent valuation approach. Am. J. Agric. Econ. 63:4.

Daubert, J. R. , R. A. Young, and S. L. Gray. 1979. Economics benefits from instream flow in a Colorado mountain stream. Washington, DC: A Report to the Office of Water Research and Technology, U.S. Dept. of the Interior.

Desvousges, W. H. , V. K. Smith, and M. P. McGivney. 1983. A comparison of alternative approaches for estimating recreation and related benefits of water quality improvements. Rep. EPA-230-05-83-001. Washington, DC: Office of Policy Analysis, U. S. Environmental Protection Agency.

Fisher, A. and R. Raucher. 1984. Intrinsic benefits of improved water quality: Conceptual and empirical perspectives. In Advances in Applied Micro-Economics, ed. V. K. Smith and D. White, pp. Greenwich, CT: JAI Press.

Freeman, A. M. 1979. The Benefits of Environmental Improvement. Baltimore, MD: Johns Hopkins University Press.

Frey, R. G. 1986. Rights, interests, desires, and beliefs. In People, Penguins, and Plastic Trees: Basic Issues in Environmental Ethics, ed. Donald Van De Veer and Christine Pierce, pp. 40–46. Belmont, CA: Wadsworth Publishing Company.

Goodin, R. E. 1980. Making moral incentives pay. Policy Sci. 12:131–145.

Goodpaster, K. 1978. On being morally considerable. J. Philosophy 75:308–325.

Graham, F. W. 1981. Cost benefit analysis under uncertainty. Am. Econ. Rev. 71:715–725.

Gramlich, F. W. 1977. The demand for clean water: The case of the Charles River. Natl. Tax J. 30:183–194.

Greenley, D. , R. Walsh, and R. Young. 1981. Option value: Empirical evidence from a case study of recreation and water quality. Q. J. Econ. 96:657–672.

Heinle, D. R. , C. F. D'Elia, J. L. Taft, J. S. Wilson, M. Cole-Jones, A. B. Caplins, and L. E. Cronin. 1980. Historical review of water quality and climatic data from Chesapeake Bay with emphasis on effects of enrichment. Pub. No. 84. Annapolis, MD: Chesapeake Research Consortium.

Henderson, J. V. and M. Tugwell. 1979. Exploitation of the lobster fishery: Some empirical results. J. Environ. Econ. Manag. 6:287–296.

Johnson, E. 1984. Treating the dirt: Environmental ethics and moral theory. In Earthbound: New Introductory Essays in Environmental Ethics, ed. Tom Regan, pp. 336–365. New York: Random House.

Kahn, J. R. and W. M. Kemp. 1985. Economic losses associated with the degradation of an ecosystem. J. Environmental Econ. Manag. 12:246–263.

Krutilla, J. 1967. Conservation reconsidered. Am. Econ. Rev. 57:777–786.

Krutilla, J. V. , C. J. Cicchetti, A. M. Freeman III, and C. S. Russell. 1972. Observations on the economics of irreplaceable assets. In Environmental Quality Analysis, ed. A. V. Kneese and B. T. Bower, Baltimore, MD: Johns Hopkins University Press.

Krutilla, J. V. and A. Fisher. 1975. The Economics of Natural Environments. Baltimore, MD: Johns Hopkins University Press for Resources for the Future.

Larkin, P. A. 1978. Fisheries management—an essay for ecologists. Ann. Rev. Ecol. Syst. 9:57–73.

Loomis, J. B. , C. F. Sorg, and D. M. Donnelly. 1986. Evaluating regional demand models for estimating recreation use and economic benefits: A case study. Water Resour. Res. 22:431–438.

McConnell, K. E. 1975. Some problems in estimating the demand for outdoor recreation. Am. J. Agric. Econ. 57:330–334.

Mäler, K. -G. 1974. Environmental Economics. Baltimore, MD: Johns Hopkins University Press.

Meyer, P. A. 1974. Recreation and preservation values associated with the salmon of the Fraser River. Inf. Rep. Ser. PAC/N-74-1. Vancouver, British Columbia, Cananda: Fish and Marine Service.

Miller, J. R. 1981. Irreversible land use and the preservation of endangered species. J. Environ. Econ. Manag. 8:19–26.

Miller, J. R. and F. C. Menz. 1979. Some economic considerations for wildlife preservation. South Econ. J. 45:718–729.

Mitchell, R. C. and R. T. Carson. 1981. An experiment in determining willingness to pay for national water quality improvements. Report to the U.S. Environmental Protection Agency. Washington, DC: Resources for the Future.

Randall, A. and J. R. Stoll. 1981. Existence value in a total valuation framework. In Managing Air Quality and Scenic Resources at National Parks and Wilderness Areas, ed. Robert D. Rowe and Lauraine G. Chestnut, Boulder, CO: Westview Press.

Regan, T. 1981. The nature and possibility of an environmental ethic. Environ. Ethics 3:19–34.

Regan, T. 1983. The Case for Animal Rights, pp. 133–147, 243–248. Berkeley, CA: University of California Press.

Riker, W. 1961. Voting and the summation of preferences. Am. Political Sci. Rev. 55:900–911.

Rodman, J. 1977. The liberation of nature. Inquiry 20:83–131.

Russell, C. B. and W. J. Vaughan. 1982. The national recreational fishing benefits of water pollution control. J. Environ. Econ. Manag. 9:328–354.

Sagoff, M. 1980. The philosopher as teacher. Metaphilosophy 11:307–325.

Sagoff, M. 1981. Economic theory and environmental law. Mich. Law Rev. 79:1393–1419.

Schulze, W. D. , D. S. Brookshire, E. G. Walther, K. K. MacFarland, M. A. Thayer, R. L. Whitworth, S. Ben-David, W. Malm, and J. Molenar. 1983. The economic benefits of preserving visibility in the national parklands of the southwest. Nat. Resour. J. 23:149–173.

Seymour, J. and H. Girardet. 1986. Far from Paradise. London: British Broadcasting Corporation.

Singer, P. 1986. Animals and the value of life. In Matters of Life and Death, 2nd ed, ed. Tom Regan, pp. 338–378. New York: Random House.

Smart, J. J. C. and Bernard Williams. 1973. Utilitarianism, For and Against. New York, N.Y.: Cambridge University Press.

Snoeyenbos, M. 1983. Business Ethics: Corporate Values in Society. Buffalo, N. Y. : Prometheus Books.

Stone, C. 1974. Should Trees Have Standing?: Toward Legal Rights for Natural Objects, pp. 1–54. Los Altos, CA: Kaufmann, William Inc.

Taylor, P. 1986. The ethics of respect for nature. In People, Penguins, and Plastic Trees: Basic Issues in Environmental Ethics, ed. Donald Van De Veer and Christine Pierce, pp. 169–184. Belmont, CA: Wadsworth Publishing Company.

Toffler, B. L. 1986. Tough Choices: Managers Talk Ethics. New York: John Wiley & Sons, Inc.

Watzlawick, P. 1976. How Real Is Real? New York: Random House.

Willig, R. D. 1976. Consumer's surplus without apology. AER 66:4.

Part Two

QUANTITATIVE APPROACHES

Chapter 8

Production Economics

MICROECONOMIC PRINCIPLES

Fixed and Variable Costs

The manager uses resources, which are the inputs, to achieve a goal or produce
a product, the output. The inputs can be fixed, or they can be variable. A
fixed input, such as a raceway, does not change, and likewise its amortized

cost does not vary over the period of amortization. In general, fixed costs are the overhead or ownership costs. Once the fixed costs are established, the manager works with variable costs. Food costs are variable and can change yearly, seasonally, or even daily. Fixed costs are important but they are often unrelated to output decisions, whereas variable costs change directly with output.

An important principle in economics and management is taken from a biological phenomenon. It is the law of diminishing returns, and applies to all culture systems; by applying it, the manager can determine the most effective (public domain) or profitable level of production. All culture systems have both fixed and variable inputs to the production process. For instance, a fixed input might be the facility, such as the number and size of ponds or the number of raceways or tanks. Another fixed input, if limited, is the available water flow or volume of water, although water flow could be a variable input if the available flow is large relative to the capacity of the facility. The common variable inputs are food and labor. The law of diminishing returns states that as units of a variable input are added to one or more fixed inputs in a culture system, the output first increases at an increasing rate, then increases at a decreasing rate, and finally decreases absolutely. In Figure 8-1 the total product (TP) line illustrates the function of this principle as it is applied in production systems.

The production function, or *TP* curve, is used to derive the average product (*AP*), as y/x, and the marginal product (*MP*), as $\Delta y/\Delta x$. The production function is then usually divided into three stages. Stage one is the portion in which the efficiency of the variable input is increasing, and *MP* is greater than *AP*. Stage

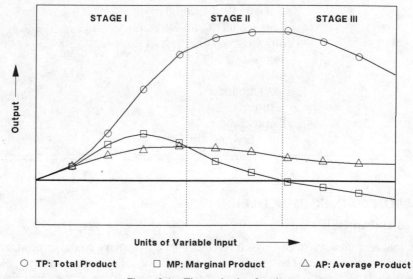

Figure 8-1. The production function.

two is the portion in which the efficiency of the variable input is declining but the marginal productivity is still positive. The most effective or profitable use of the variable input lies in stage two. Stage three is the portion of the function in which total production is decreasing, and any production in this stage is inferior in efficiency to some level in stage two.

Marginal Analysis

Suppose the manager has a production unit of fixed size, with fixed water flow, and wishes to maximize returns to that production unit using a variable food input of not more than 1 ton. Such an example can be used to illustrate the law of diminishing returns and the use of marginal analysis. Fish hatchery managers often gear production to the food conversion rate. The third column of Table 8-1 shows the food conversion rates that a manager attained in the past by feeding similar production lots. To achieve the greatest food conversion efficiency, the manager would choose to use 800 pounds of food. However, by feeding 1,000 to 1,600 pounds the manager could produce more fish, as shown in column 2. To maximize output, the manager should use 1,400 pounds of food, but by inspecting the Output column, one intuitively senses the law of diminishing returns at work. How much food should the manager use?

Maximizing Profit   Marginal analysis is a sort of shortcut method for managing the "bottom line." For instance, marginal analysis can be used to determine the level of production or amount of the variable input—food—that will maximize profits. Using the *AP* and *MP* columns in Table 8-1, we see that *AP* surpasses *MP*, or production function stage one ends and stage two begins, between 800 and 1,000 pounds of food; and *MP* goes negative, beginning stage three, at 1,600 pounds of food. The point of greatest returns, then, exceeds 800 pounds but is less than 1,600 pounds. In marginal analysis, to maximize profit, one should operate at the point at which *MC* = *MR*. Marginal cost (MC) is the price of food ($0.38/pound) divided by the marginal product (MP), where MP = change in output divided by change in input. In this example, the equality MC = MR occurs between 800 and 1,000 pounds, and by interpolation one can see that it is closer to 800 pounds. This can be confirmed by calculating actual profit, shown in the last column of Table 8.1.

Equimarginal Principle, for Multiple Products.   Maximizing profit becomes more complicated with multiple inputs and multiple outputs. Questions of allocation of scarce resources, such as dollars, deal not with added returns versus added costs, but rather with the most economical allocation to the alternatives. The equimarginal principle states that the highest total return to a scarce resource exists when the return per unit of distributed resource is equal in all alternative uses. Thus the manager should distribute the resource among enterprises or

Table 8-1. Marginal analysis table for total production (TP) output from an aquaculture unit receiving variable input of food (0–2,000 pounds)

INPUT (POUNDS OF FOOD)	OUTPUT	RATIO TO FLESH	EFFICIENCY	AP	MP	PRICE OR MR	MC	AC	VC	PROFIT
0	0	0.00	0.00	0.00	0.00	0.77	0.00	0.00	0	0
200	40	5.00	0.20	0.20	0.20	0.77	1.90	1.90	76	−45
400	170	2.35	0.42	0.42	0.65	0.77	0.58	0.89	152	−21
600	360	1.67	0.60	0.60	0.95	0.77	0.40	0.63	228	49
800	500	1.60	0.62	0.62	0.70	0.77	0.54	0.61	304	81
1000	570	1.75	0.57	0.57	0.35	0.77	1.09	0.67	380	59
1200	595	2.02	0.50	0.50	0.12	0.77	3.17	0.77	456	2
1400	600	2.33	0.43	0.43	0.02	0.77	19.00	0.89	532	−70
1600	560	2.86	0.35	0.35	−0.20	0.77	∞	1.09	608	−177
1800	500	3.60	0.28	0.28	−0.30	0.77	∞	1.37	684	−299
2000	385	5.19	0.19	0.19	−0.58	0.77	∞	1.97	760	−464

Notes: average product (AP) = Output/Input; marginal product (MP) = ΔOutput/ΔInput; marginal revenue (MR) = \$0.77/pound of product; marginal cost (MC) = \$38/100 pounds of food/MP; average cost (AC) = VC/Output; variable cost (VC) = \$0.38 × Input; Profit = (Price × Output) − VC; Fixed costs are not shown and would not affect results.

production activities, starting with the one that pays the highest net return. This principle is the basis for selecting and sizing enterprises for the optimum mix — the one that yields the greatest returns to the available resources (Meade 1985).

Least Cost Combination, for Multiple, Interchangeable Inputs. The least cost, or substitution, principle is used for decision making when there is some fixed or required level of production and some degree of substitution among the inputs. Refer to an economics text for examples of the three-step procedure using this principle.

Cost Concepts in Decision Making

To use fixed and variable costs in decision making, those costs must be considered in terms of time. Costs are classified as fixed and variable only in short-run analyses, since by definition, in the long run all costs are variable. The time value of money is discussed in Chapter 11.

TIMID COSTS OF OWNERSHIP

Taxes

If death and taxes are the only things we can count on, then the one thing in life that is an absolute is taxes. Besides entrepreneurial or corporate income taxes, there are personal property taxes on machinery and real property taxes on buildings and land. You must deal with taxes. Accept them and use the options to your advantage, starting in the early planning stages of choosing the business structure and the accounting system. According to the Internal Revenue Service, most people overpay their taxes, not due to miscalculations but simply because they do not investigate all their possible deductions and do not manage income to their tax advantage. However, not paying a tax, where and when a legal tax is due, is called *tax evasion*; it is punishable by fine or imprisonment, or both.

On the other hand, *tax avoidance* is as legal as tax evasion is illegal. Tax avoidance simply means planning your finances so that you do not pay taxes that you do not have to pay. For instance, the purchase of a tax-free municipal bond allows your money to grow tax free; that is, it allows income without a taxable dividend to report. Tax shelters were once only for the wealthy. The good news is that they are now available to everyone. The bad news is that they are now more restricted and regulated. Investigate your options thoroughly, and consider a tax advisor for the setup and strategic planning of your culture business.

For tax preparation, use the farmer's tax guide and other abundant information available from the IRS. Prepare your tax calculations early, about one month before the close of the tax year, so that you or your advisor can make credit and expenditure decisions. You may find some pleasant surprises in this presumably painful activity. Once you have prepared your estimates, consider the use of a professional tax advisor. Such "expensive" services often more than pay for themselves by saving you unnecessary tax dollars. One visit could save you money for years.

Here are some thoughts to consider on tax management. Businesses must use accrual accounting methods, but farmers may use the cash method. The cash method is simpler. One can postpone taxes by waiting to sell property after retirement. Informal income averaging can be accomplished by selling more in periods of low income and postponing sales but doing repairs and handling expense items in periods of high income. Investment tax credits are much better than depreciation. You might be able to use money you would pay in taxes to expand or improve your business.

If you are audited, be as prepared as you possibly can be; then be straightforward and polite. The auditor is doing a job that requires confronting often hostile malcontents. You and the auditor can make the process as painful or painless as you choose. If it is determined that you owe additional taxes, pay them cheerfully.

Interest

The manager should compute interest for the business on equity capital as well as on borrowed money. Computation of interest on your owned capital allows you to see and manage the opportunity cost of that capital. First, determine the average investment; then multiply that amount by the current local interest rate. For example, if a tractor has an original cost (*OC*) of $20,000, assuming a salvage value (*SV*) of $2,000 and an interest rate of 10%:

$$\text{Interest on equity} = \frac{OC + SV}{2} \times 0.10$$

$$= \frac{20,000 + 2,000}{2} \times 0.1$$

$$= \$1,100$$

Maintenance

Maintenance and repair costs are easily overlooked but are as real as any other costs. These should be budgeted as annual expenses at a rate of about 1.5% per year for new equipment and up to 5% per year for old equipment. The maintenance cost for a new $20,000 tractor then would be estimated at $300 per year.

Insurance

Insurance is a fixed cost and is usually up to 1% of the purchase price. Types of insurance are property, fire, weather, machinery, and cultured stock. Self-insuring is very risky and usually not recommended.

Depreciation

Depreciation is an accounting procedure by which a business allocates the used-up value of durable assets over the time they are owned or until salvaged. It represents a loss in value that has been used by the producer to generate income. To be depreciable, an item must be personal, tangible property or buildings used in business; have a useful life expectation of more than one year; and become useless, worn out or obsolete. Depreciable items could include pond banks and dams, pumps, casings, and the cost of drilling wells, including the costs of nonproductive wells (dry holes). Managers must look closely at the advantages of the accelerated cost recovery system (ACRS) and the modified ACRS (MACRS) of depreciation, as well as at the traditional straight line and declining balance or double declining balance methods.

The ACRS and MACRS are usually used solely for the purpose of immediate tax savings. Consult current tax forms for the accepted property classifications and recovery period percentages. The straight line method preserves more depreciation for a later period, and therefore may be a useful option for a young manager who expects incomes to increase. To calculate depreciation by the straight line method, use:

$$D_s = \frac{OC \quad SV}{L}$$

where

D_s = annual depreciation
OC = original cost (or basis)
SV = salvage value
L = expected or useful life

For a $20,000 tractor, then, the straight line method of depreciation would give:

$$D_s = \frac{20,000 - 2,000}{10}$$

$$= \$1,800/\text{year}$$

The declining balance or double declining balance method uses a rate that is double that of the straight line method. Use:

$$D_d = RV \times R$$

where

RV = remaining or undepreciated value
R = depreciation rate of twice the straight line rate

For the same tractor then, the declining balance method would give:

D_{d1} = 20,000.00 × 20% = \$ 4,000.00
D_{d2} = 16,000.00 × 20% =   3,200.00
D_{d3} = 12,800.00 × 20% =   2,560.00
D_{d4} = 10,240.00 × 20% =   2,048.00
D_{d5} = 8,192.00 × 20% =   1,638.40
D_{d6} = 6,553.60 × 20% =   1,310.72
D_{d7} = 5,242.88 × 20% =   1,048.58
D_{d8} = 4,194.30 × 20% =   838.86
D_{d9} = 3,355.44 × 20% =   671.09
D_{d10}= 2,684.35 × 20% =   536.87

$$\overline{\$17,852.52}$$

The salvage value is $20,000 minus depreciation = $2,147.48

Total Ownership Costs

The total ownership cost is the sum of the TIMID costs. For the $20,000 tractor with $L = 10$ and $SV = \$2,000$:

Taxes at 0.3%/year $\qquad = \$ \quad 60$

Interest at 10%/year $\qquad = \quad 1,100$

Maintenance at 1.5%/year $\qquad = \quad 300$

Insurance at 0.5%/year $\qquad = \quad 100$

Depreciation: $D_s = \dfrac{20,000 - 2,000}{10} = \underline{1,800}$

Total annual ownership costs $\qquad = \$3,360$

REFERENCES AND RECOMMENDED READINGS

Adams, C. M., W. L. Griffin, J. P. Nichols, and R. E. Brick. 1980. Application of a bio-economic-engineering model for shrimp mariculture systems. South. J. Agric. Econ. 135–141.

Adams, C. M., W. L. Griffin, J. P. Nichols, and R. E. Brick. 1980. Bio-engineering-economics model for shrimp mariculture systems, 1979. U.S. Sea Grant Program, TAMU-SG-80-203.

Allen, P. G., L. W. Botsford, A. M. Schuur, and W. E. Johnston. 1984. Bioeconomics of Aquaculture. New York: Elsevier Science Publishing Company.

Anderson, J. L. and J. W. Wilen. 1986. Implications of private salmon aquaculture on prices, production, and management of salmon resources. Am. J. Agric. Econ. 68:866–879.

Boyd, C. E., R. B. Rajendren, and J. Durda. 1986. Economic considerations of fish pond aeration. J. Aqua. Trop. 1:1–5.

Brown, E. E. 1977. World Fish Farming: Cultivation and Economics. Westport, CT: AVI Publishing Company, Inc.

Cauvin, D. M. and P. C. Thompson. 1977. Rainbow trout farming: An economic perspective. Can. Fish. Mar. Serv. Ind. Rep. 93.

Foster, T. H. and J. E. Waldrop. 1972. Cost-size relationships in the production of pond-raised catfish for food. Mississippi Agric. Exp. Stat. Bull. 792.

Gates, J. M., C. R. MacDonald, and B. J. Pollard. 1980. Salmon culture in water reuse systems: An economic analysis. University of Rhode Island Tech. Rep. 78.

Guerra, C. R., R. E. Risk, B. L. Godfriaux, and C. A. Stephens. 1979. Venture analysis for a proposed commercial waste heat aquaculture facility. Proc. World Maricul. Soc. 10:28–38.

Haley, K. B. 1981. Applied Operations Research in Fishing. New York: Plenum Press.

Hambrey, J. 1980. The importance of feeding, growth, and metabolism in a consideration of the economics of warm water fish culture using waste heat. In Proceedings of the World Symposium on Aquaculture in Heated Effluents and Recirculation Systems, Stavanger, May 28–30, 1980, ed. Klaus Tiews, pp. 601–617, Vol. II. Berlin: H. Heenemann.

Harsh, J. B., L. J. Connor, and G. D. Schwab. 1981. Managing the Farm Business. Englewood Cliffs, NJ: Prentice-Hall, Inc.

Israel, D. C. 1987. Comparative economic analysis of prawn hatcheries. SEAFDEC Asian Aquaculture 9:3–12.

Israel, D. C. 1987. Economic feasibility analysis of aquaculture projects: A review. SEAFDEC Asian Aquaculture, 9:6–9.

McCoy, E. W. and J. L. Boutwell. 1977. Preparation of financial budget for fish production, catfish production in areas with level land and adequate ground water. Auburn Univ. Agric. Exp. Stat. Circular 233.

Meade, J. W. 1985. Determine combinations that will maximize fish farm income. Aquaculture Mag. 11(2):27–34.

Rhodes, R. J. 1983. Primer of aquaculture finances planning for success, Part I. Aquaculture Mag. 10:16–20.

Shang, Y. C. 1974. Economic feasibility of fresh water prawn farming in Hawaii. Sea Grant Advisory Report UNIHI-SEAGRANT-AR-74-05.

Shang, Y. C. 1981. Aquaculture Economics: Basic Concepts and Methods of Analysis. Boulder, CO: Westview Press.

Shang, Y. C. 1981. A comparison of rearing costs and returns of selected herbivorous, omnivorous, and carnivorous aquatic species. Marine Fish. Rev. 43:23–24.

Shigekawa, K. J. and S. H. Logan. 1986. Economic analysis of commercial hatchery production of sturgeon. Aquaculture 51:299–312.

Sutterlin, A. M., E. B. Henderson, S. P. Merrill, R. L. Saunders, and A. A. MacKay. 1981. Salmonid rearing trials at Dear Island, New Brunswick, with some projections on economic viability. Can. Tech. Rep. Fish. Aquatic Sci. 1011. St. Andrews, NB: Department of Fisheries & Oceans.

Vondruska, J. 1976. Aquacultural Economics Bibliography. National Oceanic and Atmospheric Administration Tech. Rep. NMFS SSRF-703.

Waas, B. P., K. Strawn, M. Johns, and W. Griffin. 1983. The commercial production of mudminnows (Fundulus grandis) for live bait: A preliminary economic analysis. Tex. J. Sci. 35(1):51–60.

Waldrop, J. E. 1981. An overview of the farm-raised catfish industry in the United States—An agricultural economist's view. In Realism in Aquaculture: Achievements, Constraints, Perspectives, ed. M. Bilio, H. Rosenthal, and C. J. Sindermann, pp. 510–513. Bredine, Belgium: European Aquaculture Society.

Waldrop, J. E. and J. G. Dillard. 1985. Economics. In Channel Catfish Culture, ed. C. S. Tucker. pp. 621–645. New York: Elsevier Science Publishing Company.

Wilt, R. W. and S. C. Bell. 1977. Use of cash flow statements as a financial management tool. Auburn Univ. Agr. Exp. Stat. Bull. 487.

Yacovissi, W. A. and J. W. Meade. 1988. The economics of profit maximization. Salmonid 12(3):11–12.

Chapter 9

Records for Managerial Analyses

FOCUS: Detailed, proper, complete records are essential to
                   management.

HIGHLIGHTS: • Purposes of records
                   • What to include in records
                   • How to make an enterprise budget
                   • Inventory and valuing assets
                   • Your net worth

ORGANIZATION: Record Keeping
                   Enterprise Budgets
                   Cash Flow
                   Financial Statement

RECORD KEEPING

Records are essential, and good records can be invaluable. Records summarize operations and allow the manager to analyze the effectiveness of resource use to formulate intelligent plans. But besides measuring success and aiding planning, records are needed to comply with tax reporting laws, to obtain credit, and to make comparisons with current goals. Records provide an organized method of taking a complete inventory. They can be used to show growth, and that is important to creditors. University extension agents can provide or recommend local sources for farm-type record books. Computer software for record keeping is readily available and useful for all types of production systems, but may especially enhance the management of more complex operations. Computers allow the handling and quick retrieval of great volumes of information.

The record system should include:

Receipts
Expenses

Capital transactions
 Purchase and sale of equipment, land, and structures
 Purchase and sale of broodstock
 Depreciation schedules
Inventory
 Machinery
 Live (or cultured) stocks
 Feed and supplies
 Buildings and facility improvements
 Land
Summary of the year's business
Balance sheet (net worth statement)
Credit accounts
Labor records for Social Security
Production records

 Records that include all transactions, inventory changes, depreciation, and other pertinent information can provide a running statement of return to capital, labor, and management, as well as changes in net worth (total assets minus total liabilities), capital turnover, and other measures of business performance. Enterprise accounts are a necessary and wise investment of management efforts for optimizing the enterprise mixture and maximizing returns to resources. Managers with incomplete records tend to underestimate costs in enterprise budgets; complete records provide facts, not guesses. A system of complete records provides a tool that helps minimize errors in decision making.

ENTERPRISE BUDGETS

Budgets are written plans of action and include estimates of the expected results. Budgets are used as analytical tools for short- and intermediate-range planning. Together with cash flow analysis, budgets can provide suitable indicators of feasibility and profitability. Each enterprise or potential enterprise in a culture system can and should be scrutinized through the development of a budget. Just as a protein is formed by a combination of amino acids, the culture system is formed by an aggregate of enterprises. With the array of alternative uses for resources, as described by the enterprise budgets, the manager is in a position to review goals and resource inventories and then to undertake the all-important function of decision making. A simplified enterprise budget is presented in Table 9-1. Examples of detailed enterprise budgets, along with cash flow and credit repayment schedules, are given in Appendix III.

 Development of reliable enterprise budgets may require significant effort for what might appear to be the gathering together of little information. However, the reliability of the budget accounts for much of its value in decision

Table 9-1. Simplified enterprise budget for catfish fingerlings: Recommended management practices and estimated costs and returns per acre for 40,000 fingerlings produced from 57 broodfish, Alabama, 1983

	SIZE	UNIT	PRICE OR COST	QUANTITY	VALUE OR COST
1. Gross Receipts					
Catfish fingerlings	5.6	Inch	0.07	40,000	$2,800.00
Catfish broodfish	5.1	Pound	1.00	200	1,026.00
Total					$3,826.00
2. Variable Costs					
Broodfish	4.5	Pound	0.80	285	$1,026.00
Feed		Ton	315.00	3.5	1,102.50
Chemicals		Appl.	70.00	1.0	70.00
Fuel, oil, lubrication		Hour	2.20	13.0	28.82
Electricity		Kwh.	0.07	1400.00	19.62
Equipment (repair)		Dol.			20.00
Interest, oper. cap.		Dol.	0.14		72.47
Total					$2,339.41
3. Income Above Variable Costs					$1,486.59
4. Fixed Costs					
General overhead		Acre	5.00	1	$ 5.00
Interest on equipment		Dol.	0.12	589.70	70.76
Depreciation		Dol.			97.21
Other Machinery & Equipment costs (incl. insur., taxes)		Dol.			8.85
Total Fixed Costs					$ 181.82
5. Total Costs					$1,668.41
6. Net Returns to Land, Labor and Management					$2,157.59

making; if the budget is not realistic, it is of little value, and it may very well be of negative value. The first step in developing a budget is to set the intended magnitude of production for the enterprise, and then estimate the yield or output. Yields are usually estimated from similar operations in the local area and from information available through government and university extension specialists and agents. From the intended magnitude of production, estimates are made for the amounts and costs of required inputs. Receipts are estimated from the yield. Extension agents often have sample enterprise budgets that can be adapted to each situation. The manager should be sure to consider the average yield for the location, the level of management or cultural practices (intensification) that will be applied, and the water temperature and quality.

CASH FLOW

A cash flow statement is a summary of the culture system's receipts and expenses used to project the operation's capability to meet cash demands over a specified period of time. The statement can be used to minimize interest by helping the manager to borrow as needed rather than annually, and it provides essential information for capital investment decisions. A cash flow budget or plan, based on records from previous years, is usually prepared for a one-year period and can be used to schedule monthly inflow and outflow. A projected cash flow statement cannot be based on historical records and is less reliable and more difficult to prepare, since it must be based on estimates.

FINANCIAL STATEMENT

A financial statement consists of a listing of assets, often shown on the left-hand side of a form or sheet of paper, and a listing of liabilities, often shown beside or on the right-hand side of the form, and the "bottom line" difference between assets and liabilities, the *net worth*. Net worth is the owner's equity, or that which the owner truly owns. It is the money that would be left if all assets were sold and all credits paid. Assets usually fall into three categories: current, those used or converted to cash within one year; working, those with a two- to five-year (or intermediate) life; and fixed, or long-term, which include real estate, buildings, and improvements. Liabilities are likewise categorized and include credit for feed and supplies, and loans for equipment and mortgages. Table 9-2 is a simplified example of a financial statement. Inventories are valued in dollars by one of five methods. Of course, each method begins with a physical count or estimate of the quantity.

METHOD	EXAMPLE
Cost	Feed and supplies
Net market or selling price	Stock of fish on hand
Cost depreciation	Machinery, equipment, and new buildings
Replacement cost depreciation	Buildings over five years of age
Income capitalized value of land	Land

To determine the value of land, let:

V = capitalized value
R = $ returns to land
I = interest rate

Then:

$$V = \frac{R}{I}$$

If R = $75/acre and I = 9%:

Table 9-2. Financial statement

ASSETS		LIABILITIES	
Current		Current	
Feed and supplies	$ 11,000	Accounts Payable	$ 4,300
Fingerlings	28,000		
Fry and eggs	1,000		
Cash	2,900		
Stocks and bonds	7,600		
	$ 50,500		
Intermediate		Intermediate	
Machinery and equipment	$ 60,000	Notes	$ 29,100
Trucks	32,000		
Broodstock	9,000		
	$101,000		
Long Term		Long Term	
Buildings	$ 80,000	Mortgages	$ 91,200
Land	48,000		
	$128,000		
Total Assets	$279,500	Total Liabilities	$124,600
	−124,600		
Net Worth	$154,900		

$$V = \frac{75}{0.09}$$

$$= \$833/\text{acre}$$

REFERENCES AND RECOMMENDED READINGS

Brown, R. 1973. Mill records are part of quality control. Feedstuffs.

Chaston, I. 1984. Business Management in Fisheries and Aquaculture. Farnham, England: Fishing News Book, Ltd.

Dick, L. 1983. Agricomp software review: Transactions. AgriComp 1:12–14,46.

Hickel, R., W. Killcreas, and J. E. Waldrop. 1983. A Catfish Growth Simulation Model for Use with TRS-80 Models II, III, 16 and IBM PC Microcomputers. Agricult. Econ. Tech. Pub. No. 42, Mississippi State, MS: Mississippi State University.

Killcreas, W., S. Ishee, N. Wilkes, D. McWilliams, J. Leng, W. Wolfe, and J. Waldrop. 1985. A Records Program for Catfish and Shrimp Production; Financial Data and Management Decisions for IBM PC and Compatible Microcomputers. Agricult. Econ. Tech. Pub. No. 55. Mississippi State, MS: Mississippi State University.

Killcreas, W., S. Ishee, N. Wilkes, N. Kennedy, and E. Walker. 1985. MSU Farm Records: Summarized Capabilities, Installation, Operation and Example Reports. Agricult. Econ. Tech. Pub. No. 54. Mississippi State, MS: Mississippi State University.

Piper, R. G., I. B. McElwain, L. E. Orme, J. P. McCraren, L. G. Fowler, and J. R. Leonard. 1982. Fish Hatchery Management. Washington, DC: U.S. Fish and Wildlife Service.

Pomeroy, R. S., D. B. Luke, and J. Whetstone. 1985. Budgets and Cash Flow Statements for South Carolina Crawfish Production. Clemson University EER 83. Clemson, SC: Clemson University.

Pomeroy, R. S., D. B. Luke, and T. Schwedler. 1986. Budgets and Cash Flow Statements for South Carolina Catfish Production. Clemson University Agric. Econ. and Rural Soc. WP 052686. Clemson, SC: Clemson University.

Shaw, S. A. and J. F. Muir. 1987. Salmon: Economics and Marketing. Portland, OR: Timber Press.

Williams, C. 1983. Simple Economics and Bookkeeping for Fish Farmers. FAO Fisheries Circular No. 763 (FIRI/C763). Rome: Food and Agriculture Organization of the United Nations.

Wilt, R. W. and S. C. Bell. 1977. Use of Cash Flow Statements as a Financial Management Tool. Auburn Univ. Agric. Exp. Stat. Bull. 487.

Chapter 10

Production System Limits

FOCUS:	The determination of limits is essential for the efficient and rational management of culture systems.
HIGHLIGHTS:	• Limiting factor concept • DO and metabolite limitations • Traditional carrying capacity calculations • Bioassay procedure for your system

ORGANIZATION:	Capacity Estimates
	  A Fairyfish Tail
	  Dissolved Oxygen and Metabolites
	Production Capacity Assessment (PCA)
	  Detailed Procedure for PCA

CAPACITY ESTIMATES

"A man should know his limits."
        Clint Eastwood

The concept of limiting factors is important in production system design and in production planning. Limits exist in every aspect of every system; if they did not, we could feed the world from an aquarium. Removal of the first limiting factor generally results in increased production or production capability. As production increases the next limiting factor comes into play, and with each new limiting factor the production function follows the law of diminishing returns (described in Chapter 8) as units of variable input are added. To use resources efficiently, the manager must find the limits of the system.

You can use production capacity limits in planning and scheduling as both health management guidelines and production incentives. Although limits are often set by design, for economic or organizational reasons, the physical and chemical limits of the system should also be identified. Limits are determined by experience (trial and error), by experiment, or through calculations based on scientifically sound data. Limits may be established, but not determined, by opinion and administrative fiat. Often erroneous conclusions, drawn from data or from conclusions of others about the meaning of data, are used as, and confused with, actual physical and biological limits. Failure to establish limits is foolish if not dangerous. (What did Clint Eastwood say?) True limits are unforgiving, and surpassing them can be catastrophic. On the other hand, a policy of never approaching the limits is a safe, low-risk strategy—and one that can be horribly inefficient.

A Fairyfish Tail

Once upon a time, it was accepted that 1 acre of sport fish pond would support 500 pounds of fish. Then one day three clever young men tried to increase the weight of fish in a pond—and lo and behold, they harvested 1,500 pounds of fish per acre the next year. They were very proud. After some time, they decided to take a chance and push the weight toward nearly 1 ton per acre. The edging-up process continued until one day, about 30 years later, those clever old men were able to rear over 5,000 pounds of fish per acre. They had not changed their technology much, but they had many more ponds and, most importantly, they had 10 times as many fish per acre. And they were proud to say that they had never lost a fish because they had surpassed the limit of a pond.

One day a friend innocently asked if they could have tried a higher density, like the 5,000 pounds per acre, in one small pond 20 years ago. (Could they, without knowing it, have produced at that higher rate throughout their careers? Could they have made a lot more money, produced far more sport fish at a cheaper price for anglers, taken their wives on a nice vacation, and saved 20 years of production evolution?) After a momentary silence, the three old men took turns saying "no" several times while frowning and shaking their heads. (After all, they each had had 30 years of experience.) The friend said that was good to hear, and good to know that they had not wasted most of their (ponds') production potential for most of their lives. All of a sudden, one of the angry old men picked up a bucket of rotten fish goo and poured it on the former friend's shoes. Another told the former friend that he was mean and should leave them alone. The last sad old man just sighed and said a bad word. Figure 10-1 illustrates the potentially lost production.

Dissolved Oxygen and Metabolites

Production of aquatic animals is usually limited first by dissolved oxygen (DO). Other limiting factors may include total metabolites (excreted by-products of digestion), ammonia or un-ionized ammonia, water treatment or handling capabilities, space, solids (total, dissolved, suspended), temperature, food, and more. The terms used to describe the intensity and limits of fish production in flowing-water culture systems are often ill-defined and unclear. However, Banks and Fowler (1982) and Westers (1983) explained the differences in the density concept as weight per volume and in the loading concept as weight per flow.

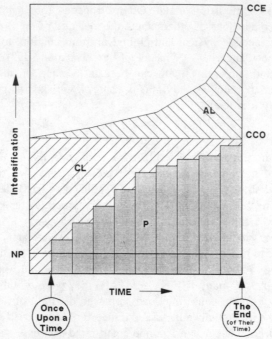

NP – Natural Production
P   – Production due to management of clever young men
CCO – Carrying Capacity based on technology potentially
        available Once Upon A Time
CCE – Carrying Capacity based on technology potentially
        available at the End of Their Time
CL – Conspicuous lost production
AL – Additional lost production

Figure 10–1. Fish production per acre of pond, achieved by some clever young men.

Piper et al. (1982) defined carrying capacity as the animal load a system can support, stated as pounds of fish per cubic foot (lb/ft^3) of water, although they noted that reference is often made to flow as well as volume. Haskell (1955) made two assumptions regarding carrying capacity. First, carrying capacity is limited by the rate of oxygen consumption and by the accumulation of metabolites; second, those rates are proportional to the amount of diet fed per day. By trial and error, Haskell determined a carrying capacity that he believed was reasonable for brown trout, *Salmo trutta*, in New York State fish hatcheries. He based the relation on pounds of diet per day per cubic foot of water, which was 0.13 lb/ft^3 for raceways and 0.02 lb/ft^3 for ponds. He did not consider variation in flow rate but simply flow (raceway) versus static (pond) situations. Haskell's calculation was necessarily limited by the weight/volume factors; however, once feeding requirements were known for a fish fed a particular diet (any feeding chart is suitable), the following equation could be used:

$$\text{Weight of diet} = \frac{\%BW \times \text{ total weight of fish}}{100}$$

Thus

$$\text{Weight of fish} = \frac{\text{weight of diet} \times 100}{\%BW}$$

where BW = body weight

and production could be estimated by:

$$\text{Gain} = \frac{\text{weight of diet}}{\text{conversion rate}}$$

where conversion rate = weight of food/weight gain of fish.

Later, Haskell (1959) concluded that trout length increases at a constant rate (absolute increment per unit time) at a particular temperature, and body form remains constant for at least one and a half years. Haskell also concluded that when a pond held a maximum amount of trout, the amount of food required to feed those fish would always remain the same, regardless of species, size, or temperature. "Hence, that amount may be considered . . . its carrying capacity." (It has been established elsewhere, however, that temperature can have a major effect on carrying capacity, and the oxygen demand of metabolism will depend on the composition and digestibility of the particular diet for the fish being reared.) Thus carrying capacity was defined in terms of the amount of diet fed. Haskell (1959) then showed that:

$$\%BW = \frac{\text{weight} \times \text{conversion} \times \Delta L \times 100}{L}$$

where:

L = length in inches
ΔL = change in length per day

Buterbaugh and Willoughby (1967) defined the numerator in Haskell's (1959) feeding rate equation as a hatchery constant (HC) for fish reared at any fixed or single temperature. They presented a feeding guide that could be used anywhere once the HC was empirically derived. They also illustrated the use of Haskell's (1959) temperature unit concept to estimate growth and thus allow the use of the hatchery constant feeding system where water temperature varies.

Burrows and Combs (1968) introduced the concept of water reuse when they stated that reconditioning of rearing water requires the replenishment of oxygen and removal of carbon dioxide and ammonia. Using chinook salmon weighing up to 17 g as an example, they presented carrying capacity graphically for six water temperatures, ranging from 40 to 65°F, as pounds of fish per gallon per minute (gpm) of water flow. Thus, the present concept of loading, or oxygen demand of biomass per unit flow of water, was introduced, albeit unnamed. Burrows and Combs also stated that fish density, or crowding, is important with regard to stress and feeding efficiency, and that fish weight per cubic foot could be safely increased two- to threefold as fish grew from 1 to 50 g.

Willoughby (1968) developed a calculation from an unpublished hypothesis of Haskell (Harry Westers, personal communication) which stated that carrying capacity, based on amount of diet per day, was proportional to rearing unit water exchanges per hour. Willoughby's method predetermined rearing unit carrying capacities without the trial-and-error values necessary for earlier calculations. Also, it was not constrained by empirical data for any species, as in the example of Burrows and Combs (1968), although it was based on trout diets and required an estimate for oxygen consumption. Willoughby showed that available oxygen, as incoming minus outgoing concentration (usually a minimum concentration set by the fish culturist), could be used to determine the daily food allotment. He calculated the 24-hour oxygen availability as (O_{in} − O_{out}) × 0.0545 × gpm, where:

$$0.0545 = \frac{\text{tons of water in 24 hours/gpm}}{\text{g of } O_2 \text{ to metabolize 1 lb of trout diet}}$$

His *formula*, as it is often referred to by fish culturists, equated the 24-hour oxygen availability to pounds of diet per day.

Westers (1970) introduced the concept of rearing unit exchange rate when he presented a series of five graphs stating the maximum number of pounds of coho salmon, weighing 1 to 43 g, which could be supported by a water flow of one to six exchanges per hour. Each graph illustrated the carrying capacity for a temperature range of 5°F, and all graphs covered the range of 40 to 65°F. Westers' equation for the carrying capacity of a rearing unit was

$$\text{lb/gpm} = \frac{\text{lb/ft}^3 \times 8}{R}$$

where:

R = exchanges per hour
8 = constant conversion factor,   (60 gal/hr)/(7.48 gal/ft^3)

Westers and Pratt (1977) stated that water quality determines ultimate production capacity, and they used concentrations of DO and ammonia to set biological limits for production, expressed as weight of fish per unit of flow (kg per L/min). Through the use of two figures and two graphs, one could determine the carrying capacity of any suitable water source, assuming 0.0125 mg/L as the allowable un-ionized ammonia concentration. The implicit assumption was that aeration would resupply oxygen, if needed, until the allowable ammonia level was reached.

Several other equations have been developed to calculate carrying capacities (Table 10-1). A number of articles present useful reviews or present the carrying capacity concept through additional methods of calculation or estimation, including Seece (1973), Piper (1970), Buss (1982), and Westers (1978 and 1981).

Table 10-1. Reported methods of carrying capacity calculation for intensive fish culture systems in North America

REFERENCE	CALCULATION(C)/INTERPRETATION (I)	
Haskell (1955)	C	$lb/ft^3 = \frac{0.13 \times 100}{\% BW}$
	I	Brown trout density per cubic foot of raceway; use New York State feed chart; flow not considered. BW = body weight
Haskell (1959)	C	$lb/gpm = \frac{max\ lb\ diet}{\% BW}$
	I	Capacity of a unit for any size and species is weight of fish which could be fed diet amount found to be maximum for the unit.
	C	$\% BW = \frac{3 \times conversion \times \Delta L \times 100}{L}$
	I	Diet amount per day as $\% BW$ of fish
Buterbaugh and Willoughby (1967)	C	$\% BW = \frac{HC}{L}$
	I	$HC = 300 \times conversion \times \Delta L$. Use with authors' feed chart.
Burrows and Combs (1968)	C	Extrapolate determination from graph.
	I	Pounds of chinook/gpm and cubic feet, by size
Willoughby (1968)	C	$lb\ diet/day = (O_{in} - O_{out}) \times 0.0545 \times gpm$
	I	Allowable trout diet (1.2 kcal/lb) to feed per day. O = oxygen concentration in mg/L

Table 10-1 (*cont.*)

REFERENCE	CALCULATION(C)/INTERPRETATION (I)	
Westers (1970)	C	$\text{lb/gpm} = \frac{\text{lb/ft} \times 8}{R}$
	I	R = hourly exchanges
Liao (1971)	C	$Q = \frac{1.2(C_e - C)}{O}$
	I	Q = lb fish/gpm; $C_e - C$ = available DO in mg/L; O = oxygen uptake rate
Buss and Miller (1971)	C	$Q = 37.41W \times D$
	I	Q = gpm; W = width in ft; D = depth in ft
	C	$L_t = 0.667P/W \times D$
	I	L_t = length of unit; P = lbs of production
Westers and Pratt (1977)	C	$\frac{\text{kg/L}}{\text{min}} = \frac{W \times L \times uses}{T}/\text{L/min}$
	I	W = weight of 100-mm fish at a specific HC; L = fish length in mm; $T = {}^\circ C$
	C	$\text{kg/m}^3 = \frac{\text{kg/lpm}}{(R/0.06)}$
	I	R = hourly exchanges

PRODUCTION CAPACITY ASSESSMENT (PCA)

You can use bioassays to set actual, on-site production limits. Before you develop a water supply, you will probably want to ensure that fish live or have lived in that supply—even if the chemistry profile looks good. Similarly, before scheduling production and establishing a precedent for intensification at the facility, you should physically test the limits if there is concern that any factor other than oxygen could (even if only for a short time) become limiting. If you plan to use an oxygenation system, you must either depend on calculations based on un-ionized ammonia or conduct a trial (a bioassay) to determine the safe limits for production.

The methods used for acute bioassay generally determine the concentration at which 50% of the fish are killed within a specified time, such as 96 hours. This lethal concentration, the $LC\text{-}50_{96}$, is then divided by 20, or at least by 10, to provide an environmentally safe standard. A major problem with the acute method is that the physiological mechanisms involved with mortality may be entirely different from the mechanisms affected in a chronic, or lower-concentration, situation. Thus the data on acute toxicity may provide little if any information that will establish safe chronic-effects levels.

The production capacity assessment (PCA) bioassay is one method used to determine the effects of a gradient of culture intensification. The PCA procedure is suitable for flowing water (raceway) systems but is not designed for pond production assessment. A stepwise series of small rearing units, such as tubs, tanks, or aquaria, is set up on the production site to serially reuse (in a single pass and not to recirculate) the source water. The same fish or aquatic organisms that are to be reared in that water at that site are used to load each rearing unit sufficiently to remove 2.5 to 3 mg/L of DO from the flow through each unit. At the inlet of each assay rearing unit the water is aerated, in a one-way flow, before passing into the next rearing unit. In this way the DO constraint is removed, but the fish are exposed to any buildup of metabolites or any other factor related to increasing intensification that is likely to occur in that production system. After the mean growth rate is established over several weeks for fish in each rearing unit, the growth rates are regressed against the cumulative milligrams per liter of DO removed from water at the corresponding rearing unit points in the series. The manager must choose a maximum acceptable reduction in growth. The bioassay result then allows the manager to predict from the regression line the cumulative milligrams per liter of DO that is the acceptable limit of intensification for that system. The following section provides a more detailed procedure, and Appendix VI contains sample problems that demonstrate the use of the calculation.

Detailed Procedure for PCA

Prepare a stepwise series of at least five rearing units, so that each unit after the first receives water from the previous unit. Provide an aeration chamber with one-way water flow at the inlet of each rearing unit. Stocking rate of fish in each unit should be sufficient to reduce DO by 25–30% within the unit $(DO_{in} - DO_{out} = DO_{in} \times 0.7)$. Aeration between successive units should be sufficient to restore DO to within 90% of saturation.

Scale the unit size and flow to provide at least three exchanges of water per hour, and remove settled solids daily. During removal of solids, replace any water lost to the system (by additional inflow at a rate roughly equal to the water loss during cleaning) to ensure that at least the minimum flow is maintained to all units at all times. A trial is used to establish flow and adequacy of aeration. Fish are added to one unit until the DO is reduced by 25%. Maintain the flow and fish for at least 24 hours while you monitor ΔDO across the tank and across the following aeration chamber. Operate all aerators during the trial even though only one is functional. Make adjustments to flow, fish, and aeration as needed.

Once the operational variables (stocking rate, flow, and aeration) are established and stable, remove the fish from the trial unit and discard them. Increase the flow to roughly 300% of the flow established in the trial. Add fish to all units in a random stocking fashion. Reduce the flow over a three-day

Table 10-2. Data for PCA determination

SERIAL USE	START WT (KG)	END WT (KG)	SGR	$DO_{in} - DO_{out}$ (MG/L)	CUMULATIVE ΔDO
1	4.49	5.39	0.56	2.7	2.7
2	4.50	5.27	0.49	3.2	5.9
3	4.48	4.94	0.30	2.8	8.7
4	4.37	4.65	0.19	2.4	11.1
5	3.86	3.90	0.03	2.2	13.3

Note: The number of days in this example is 14.

period; after one day $\rightarrow$ 200%; two days $\rightarrow$ 150%; three days $\rightarrow$ 100% of established flow.

Feeding should be full rations as planned for production. During the three-day start-up period, however, feeding should be at half of the full ration rate. After two weeks, weigh the fish and use this weight as the starting weight for growth determination. After two additional weeks, during which time oxygen consumption ($DO_{in} - DO_{out}$) is monitored before feeding, reweigh the fish. If there were no major problems, the test is complete. If the manager desires, the test can be continued for another two to four weeks to establish the growth and oxygen consumption rates more precisely. The test should end before any disease outbreak occurs or, if disease is encountered, when mortality reaches 0.2% per day in any rearing unit.

Arrange the data, as in Table 10-2, to provide the cumulative oxygen consumption and specific growth rate (SGR), as (log Wt_{end} − log Wt_{start}/days) × 100. The linear regression for the effective cumulative oxygen consumption at which growth is reduced by some fixed percentage, say 50%, of maximum ($ECOC_{50}$ in milligrams of DO per liter) is accomplished by using the cumulative ΔDO as the x variable and SGR as the y variable. Using 0.28 as the 50% of growth point for prediction of allowable DO removal, the result (r = 0.985), or the $ECOC_{50}$, is 9.0 mg DO/L.

REFERENCES AND RECOMMENDED READINGS

Adams, C. M., W. L. Griffin, J. P. Nichols, and R. E. Brick. 1980. Application of a bio-economic-engineering model for shrimp mariculture systems. South. J. Agric. Econ. 135–141.

Adams, C. M., W. L. Griffin, J. P. Nichols, and R. E. Brick. 1980. Bio-engineering-economics model for shrimp mariculture systems, 1979. U.S. Sea Grant Program, TAMU-SG-80-203.

Allen, P. G., L. W. Botsford, A. M. Schuur, and W. E. Johnston. 1984. Bioeconomics of Aquaculture. New York: Elsevier Science Publishing Company.

Banks, J. and L. G. Fowler. 1982. The effects of population weight loads and crowding on fall chinook fingerlings reared in circular tanks. Technical Transfer Series No. 82–3. Washington DC: U.S. Fish and Wildlife Service.

Boyd, C. E., R. B. Rajendren, and J. Durda. 1986. Economic considerations of fish pond aeration. J. Aqua. Trop. 1:1–5.

Brown, E. E. 1977. World Fish Farming: Cultivation and Economics. Westport, CT: AVI Publishing Company, Inc.

Burrows, R. E. and B. D. Combs. 1968. Controlled environments for salmon propagation. Prog. Fish-Cult. 30(3):123–136.

Buss, K. 1982. Calculating the productivity of a water supply for aquaculture. In Proceedings of the Commercial Trout Farming Symposium. Pages 86–91. Reading, England: The Institute of Fisheries Management, Reading University.

Buss, K. and E. R. Miller. 1971. Considerations for conventional trout hatchery design and construction in Pennsylvania. Prog. Fish-Cult. 33:86–94

Buterbaugh, G. L. and H. Willoughby. 1967. A feeding guide for brook, brown, and rainbow trout. Prog. Fish-Cult. 29:210–215.

Cauvin, D. M. and P. C. Thompson. 1977. Rainbow trout farming: An economic perspective. Can. Fish. Mar. Serv. Ind. Rep. 93.

Couper, J. R. and W. H. Rader. 1986. Applied Finance and Economic Analysis for Scientists and Engineers. New York: Van Nostrand Reinhold Company.

Foster, T. H. and J. E. Waldrop. 1972. Cost–size relationships in the production of pond-raised catfish for food. Mississippi Agric. Exp. Stat. Bull. 792.

Gates, J. M., C. R. MacDonald, and B. J. Pollard. 1980. Salmon culture in water reuse systems: An economic analysis. Univ. of Rhode Island Tech. Rep. 78.

Guerra, C. R., R. E. Risk, B. L. Godfriaux, and C. A. Stephens. 1979. Venture analysis for a proposed commercial waste heat aquaculture facility. Proc. World Maricul. Soc. 10:28–38.

Haley, K. B. 1981. Applied Operations Research in Fishing. New York: Plenum Press.

Hambrey, J. 1980. The importance of feeding, growth, and metabolism in a consideration of the economics of warm water fish culture using waste heat. In Proceedings of the World Symposium on Aquaculture in Heated Effluents and Recirculation Systems, Stavanger, May 28–30, 1980, ed. Klaus Tiews, pp. 601–617 Vol. II. Berlin: H. Heenemann.

Haskell, D. L. 1955. Weight of fish per cubic foot of water in hatchery troughs and ponds. Prog. Fish-Cult. 17(3):117–118.

Haskell, D. L. 1959. Trout growth in hatcheries. N.Y. Fish Game J. 6:205–237.

Israel, D. C. 1987. Comparative economic analysis of prawn hatcheries. SEAFDEC Asian Aquaculture 9:3–12.

Israel, D. C. 1987. Economic feasibility analysis of aquaculture projects: A review. SEAFDEC Asian Aquaculture 9:6–9.

Larkin, P. A. 1988. The future of fisheries management: Managing the fisherman. Fisheries 13(1):3–9.

Liao, P. B. 1971. Water requirements of salmonids. Prog. Fish-Cult. 33:210–215.

McCoy, E. W. and J. L. Boutwell. 1977. Preparation of financial budget for fish production, catfish production in areas with level land and adequate ground water. Auburn Univ. Agr. Exp. Stat. Circular 233.

Meade, J. W. 1988. A bioassay for production capacity assessment. Aquacult. Eng. 7:139–146.

Piper R. G. 1970. Know the proper carrying capacity of your farm. Am. Fish. U.S. Trout News 15:4–6, 30.

Piper R. G., I. B. McElwain, L. E. Orme, J. P. McCraren, L. G. Fowler, and J. R. Leonard. 1982. Fish hatchery management. Washington DC: U.S. Fish and Wildlife Service.

Shang, Y. C. 1974. Economic feasibility of fresh water prawn farming in Hawaii. Sea Grant Advisory Report UNIHI-SEAGRANT-AR-74-05.

Shang, Y. C. 1981. A comparison of rearing costs and returns of selected herbivorous, omnivorous, and carnivorous aquatic species. Mar. Fish. Rev. 43:23–24.

Shigekawa, K. J. and S. H. Logan. 1986. Economic analysis of commercial hatchery production of sturgeon. Aquaculture 51:299–312.

Speece, R. E. 1973. Trout metabolism characteristics and the rational design of nitrification facilities for water reuse in hatcheries. Trans. Am. Fish. Soc. 102:323–334.

Sutterlin, A. M., E. B. Henderson, S. P. Merrill, R. L. Saunders, and A. A. MacKay. 1981. Salmonid rearing trials at Dear Island, New Brunswick, with some projections on economic viability. Can. Tech. Rept. of Fisheries and Aqu. Sci. 1011. St. Andrews NB: Department of Fisheries and Oceans.

Vondruska, J. 1976. Aquacultural Economics Bibliography. National Oceanic and Atmospheric Administration Tech. Rep. NMFS SSRF-703.

Waas, B. P., K. Strawn, M. Johns, and W. Griffin. 1983. The commercial production of mudminnows (*Fundulus grandis*) for live bait: A preliminary economic analysis. Tex. J. Sci. 35:51–60.

Waldrop, J. E. 1981. An overview of the farm-raised catfish industry in the United States—An agricultural economist's view. *In* Realism in Aquaculture: Achievements, Constraints, Perspectives, ed. M. Bilio, H. Rosenthal, and C. J. Sindermann, pp. 519–533. Bredine, Belgium: European Aquaculture Society.

Westers, H. 1970. Carrying capacity of salmonid hatcheries. Prog. Fish-Cult. 32:43–46.

Westers, H. 1978. Biological considerations in hatchery design for coolwater fishes. *In* Selected Coolwater Fishes of North America, ed. R. L. Kendall, pp. 246–253. Am. Fish. Soc. Spec. Pub. 11. Washington DC: American Fisheries Society.

Westers, H. 1981. Fish Culture Manual for the State of Michigan (Principles of Intensive Fish-Culture). Lansing, MI: Michigan Department of National Resources.

Westers, H. 1983. Considerations in hatchery design for the prevention of diseases. *In* A guide to integrated fish health management in the Great Lakes basin, eds. F. P. Meyer, J. W. Warren, and T. G. Carey, pp. 29–35. Ann Arbor, MI: Great Lakes Fishery Commission.

Westers, H., and K. M. Pratt. 1977. Rational design of hatcheries for intensive salmonid culture based on metabolic characteristics. Prog. Fish-Cult. 39:157–165.

Willoughby, H. 1968. A method for calculating carrying capacities of hatchery troughs and ponds. Prog. Fish-Cult. 30:173–174.

Chapter 11

Decision-Making Tools

FOCUS: Tools don't make decisions for the manager;
               however, there are techniques that often can be
               used to help identify, gather, organize and analyze
               the most important criteria.

HIGHLIGHTS:
- Use partial budgets for a quick bottom-line determination
- A forecasting technique (Delphi)
- Foundations for economic analysis
- Discounting for comparison of future events
- Table of "When to Use" analysis techniques

ORGANIZATION: Partial Budgeting
                   The Delphi Technique
                   Benefit-Cost Analysis
                     Present Value Analysis
                     Five Additional Analysis Techniques
                       Benefit/Cost Ratio
                       Uniform Annual Cost
                       Savings/Investment Ratio
                       Discounted Payback Analysis
                       Break-Even Analysis
                     Sensitivity Analysis
                   Decision Trees

Decision-making tools or decision support systems do not replace the manager's responsibility for ensuring that a decision is made; rather, they do exactly what you tell them to do. The decision-making process involves intelligence, design, and choice. Intelligence is the identification and definition of the problem and all the reasonable alternative solutions to it. Design is the identification of one or more suitable methods to evaluate the alternatives and the conduct of the evaluations. Choice is your stated commitment to an alternative or your

111

selection of the best alternative in terms of a stated objective such as profitability or effectiveness.

PARTIAL BUDGETING

Partial budgeting is used with an enterprise budget or an entire culture system budget to evaluate a contemplated change. Partial budgeting is especially useful for relatively small operations. Once the enterprise budgets are well prepared, a partial budget is a useful tool to help the manager decide quickly whether to change an enterprise, purchase or rent machinery, grow out yearlings, or sell fingerlings. Partial budgeting indicates whether the change a manager is considering is more or less cost effective than the status quo. The analysis addresses four items for the alternative—cost reductions, cost additions, return reductions, and return additions—and arrives at the bottom line or net difference.

To prepare a partial budget, inspect the enterprise (or culture system) budget and list the amount of the expected change for each of the four categories. Add up all additional returns, plus the amounts of all reductions in costs. From that sum, subtract all additional costs and the amounts of all reduced returns. The partial budget is then complete. If the answer is positive, the change will result in a net increase by the amount shown. If the answer is negative, the change will result in a net loss to the operation.

THE DELPHI TECHNIQUE

Ah, if we could only predict the future, we would all be wealthy stock speculators. The oracle at Delphi dictated the future; the best we can do is to make an educated guess. One way to stack the deck in our favor might be to ask the advice of experts. But what happens if their advice is highly variable (as it almost assuredly will be)? One could bring them together in a meeting and see from the interactions which advice has the greatest support of the group. But if there is a leader or teacher–student relation or strong and weak personalities involved, the group might be swayed for the wrong reasons.

Or one could go back to each expert separately, with a summary of the inputs of the as yet unidentified experts, and see if that summary information might help the individual expert to be more precise and to give an answer that is closer to the group mean. And one could summarize that second round of information, again take it to each expert, and perhaps come even closer to the most likely answer based on all the expert opinions. In conducting such a procedure, the manager would be applying the Delphi method or technique.

BENEFIT-COST ANALYSIS

The benefit-cost concept may be abused more often than it is properly used. There are six traditional techniques or alternatives in benefit-cost analysis. However, before considering them, it is important to review the foundation for economic analyses in general, along with the steps one should follow in developing and conducting an analysis. The following example demonstrates a use of benefit-cost analysis.

REARING UNIT USE	A	B
Alternatives	Fingerlings	Yearlings
Benefits	$10,000 (earned)	$15,000 (earned)
Costs	Feed and labor = $1,000	Feed and labor = $2,000
Benefit/cost ratio	$10.00 (or 10:1)	$7.50 (or 7.5:1)

Your choice? Anyone using a traditional benefit-cost analysis can tell that you should choose alternative A. And anyone can lose $4,000 per raceway.

An economic analysis is a systematic approach for comparing the alternatives to reaching an objective. All reasonable alternatives should be considered, and the current and future costs of each must be included in a way that accounts for the time value of money. The analysis itself will not set priorities among goals or objectives, nor is the analysis a decision-making process. It is merely a systematic way of looking at inputs and the resulting outputs that indicates the most cost-effective option based on quantifiable, usually monetary, values. It cannot address the many important qualitative considerations such as health, safety, morale, or politics. Naturally, the outputs can be no better than the inputs; thus, it is critical to formulate assumptions carefully and to estimate costs and benefits precisely. As in all estimates and assumptions, the conclusions are inherently uncertain.

All benefit-cost analyses must include at least three steps: identification of alternatives, determination of the costs and benefits for each alternative, and a comparison of the costs and benefits of all alternatives. A more completely defined set of steps, although all are not required for every analysis, is:

1. Define the objective; it must be specific, detailed, and attainable.
2. Formulate assumptions.
3. Identify alternatives.
4. Determine the costs and benefits of each alternative.
5. Evaluate the alternatives by comparing their costs and benefits.
6. Test the outcome sensitivity to major uncertainties.
7. Present (report) the synthesized results.
8. Recommend an alternative.

Although assumption formulation is listed as the second step, it is important to realize that the entire analysis is infused with assumptions. Certain assumptions are associated with each of the six economic analysis techniques. Assumptions should be formulated only to bridge gaps due to identified present and future uncertainties. Assumptions should be distinguishable from facts, they should be realistic, and they must be essential. Typical examples of assumptions used in nearly all benefit-cost analyses include the estimate of future resources use, the economic life of a project, and the period of comparison.

Costs generally fall into three categories. Sunk costs are those already expended and are *not* included in an economic analysis. Nonrecurring costs include such things as construction and purchases of long-life-expectancy items. Recurring costs are exemplified by the operating costs associated with labor and materials.

Once the manager has considered and addressed all the preliminaries just mentioned, established the objective, identified alternatives, recognized assumptions, and estimated the costs and benefits of each alternative, it is time to compare the alternatives. The first technique to be considered, present value analysis, must be used in every benefit-cost analysis when the economic life exceeds three years, and should be used for all projects regardless of their economic life. The analysis simply values all life-cycle costs and benefits at their present worth.

Present Value Analysis

Usually all factors to be considered in an analysis do not favor the same alternative. It is then necessary to put all the alternatives on a common basis of time and cost to make a valid comparison. The simplest way to compare the alternatives is to perform a present value (PV) analysis. The PV is the estimated current worth of future benefits or costs derived by discounting the future values, using an appropriate discount rate. The alternative having the lowest PV cost is considered the least costly alternative over the life of the project.

To use PV analysis as the sole basis for decision making, the following assumptions apply:

1. Benefits that have not been put in monetary terms are equal for all alternatives. When benefits are not equal, the least costly alternative is not necessarily the best alternative. The best alternative may in fact be the one that costs more, yet produces a significantly higher level of benefits. Thus, when benefits are unequal, the decision should not be based solely on the PV analysis.
2. The intended estimated life of the alternative must be specified.
3. Service lives of alternatives must either be equal or be placed on an equal basis. This can be accomplished by using the common multiple approach, or by comparison using the shorter life and inputing the residual value of the asset with the longer life.

A standard discount rate must be used in evaluating the measurable costs and benefits of alternatives when they are distributed in the future. This discount rate is used to derive discount factors for each month. The discount factors are then multiplied by the monthly cost or benefit to determine the PV cost or benefit. Monthly cost or benefit × discount factor = PV cost or benefit. Appendix IV contains details on life-cycle costing methods.

To calculate the PV for each alternative, determine the costs and the future month in which the costs will be incurred. Add all costs within each month to determine total costs for that month. Look up the monthly discount factor for each month (Table 11-1 gives discount factors based on a 10% rate) and multiply that factor by the total monthly cost to determine PV costs for each month. Add up all PV costs.

Table 11-1. Project month discount factors based on an annual rate of 10%

MONTH	FACTOR	MONTH	FACTOR	MONTH	FACTOR
1	.992 089	37	.745 371	73	.560 008
2	.984 240	38	.739 474	74	.555 578
3	.976 454	39	.733 624	75	.551 183
4	.968 729	40	.727 821	76	.546 822
5	.961 066	41	.722 063	77	.542 496
6	.953 463	42	.716 351	78	.538 205
7	.945 920	43	.710 683	79	.533 947
8	.938 436	44	.705 061	80	.529 723
9	.931 012	45	.699 483	81	.525 532
10	.923 647	46	.693 950	82	.521 375
11	.916 340	47	.688 460	83	.517 250
12	.909 091	48	.683 013	84	.513 158
13	.901 899	49	.677 610	85	.509 098
14	.894 764	50	.672 249	86	.505 071
15	.887 686	51	.666 931	87	.501 075
16	.880 663	52	.661 655	88	.497 111
17	.873 696	53	.656 421	89	.493 179
18	.866 784	54	.651 228	90	.489 277
19	.859 927	55	.646 076	91	.485 406
20	.853 124	56	.640 965	92	.481 566
21	.846 375	57	.635 894	93	.477 757
22	.839 679	58	.630 863	94	.473 977
23	.833 036	59	.625 873	95	.470 227
24	.826 446	60	.620 921	96	.466 507
25	.819 908	61	.616 009		
26	.813 422	62	.611 136		
27	.806 987	63	.606 301		
28	.800 603	64	.601 505		
29	.794 269	65	.596 746		
30	.787 986	66	.592 025		
31	.781 752	67	.587 342		
32	.775 567	68	.582 695		
33	.769 432	69	.578 085		
34	.763 345	70	.573 512		
35	.757 306	71	.568 975		
36	.751 315	72	.564 474		

As an example (adapted from USDOT 1985), a fish processing cooperative currently has its automated data processing (ADP) needs met through a time-sharing agreement. Two alternatives are being considered as replacements for the time-sharing agreement. Alternative one is to purchase a mainframe system for $90,000; alternative two is to rent the mainframe system for $2,000 per month. Other costs are detailed below. The system has an economic life of six years starting in October.

	STATUS QUO	BUY	RENT
Nonrecurring			
Purchase	—	$90,000	—
Site/facility	—	$20,000	$20,000
Construction			
Recurring			
Rent	—	—	$2,000/mo
Time sharing	$3,500/mo	—	—
Maintenance	—	$175/mo	$175/mo
Personnel			
Salaries	$2,000/mo	$2,750/mo	$2,750/mo
Training	—	$1,200 each October	$1,200 each October
Equipment			
Software	—	$500 each October	$500 each October
Supplies	$150 the first month	$150 the first month	$150 the first month
	$55 per month thereafter	$55 per month thereafter	$55 per month thereafter

From the total costs we see that in today's money it would be cheaper to buy. If we had to pay (by establishing a trust fund) for all future costs today, we would need $280,044 to buy, $299,990 to rent, or $303,475 to continue as is for the next six years.

SITE:_____

START OF ANALYSIS PERIOD:_____

LENGTH OF ANALYSIS PERIOD:_____

	STATUS QUO		BUY		RENT	
	CONSTANT	PV	CONSTANT	PV	CONSTANT	PV
FY 1988	66,755	63,410	147,555	144,867	81,555	78,383
FY 1989	66,660	57,578	37,460	32,421	61,460	53,151
FY 1990	66,660	52,334	37,460	29,469	61,460	48,311
FY 1991	66,660	47,543	37,460	26,787	61,460	43,915
FY 1992	66,660	43,257	37,460	24,358	61,460	39,932
FY 1993	66,660	39,324	37,460	22,143	61,460	36,301
Total						
Costs	400,055	303,475	334,855	280,044	388,855	299,992

Five Additional Analysis Techniques

If further evaluation is needed, five additional techniques may be used (Table 11-2): benefit/cost ratio, uniform annual cost, savings/investment ratio, discounted payback analysis, and break-even analysis. The benefit/cost ratio is the only analysis that is useful when the benefits from alternatives are unequal. The break-even analysis is popular because the associated graphs are straightforward, easily understood, and allow the manager to compare the alternatives visually at any time (Fig. 11-1). The following section, taken from USDOT (1985), contains details on the use of each analysis technique. Table 11-3 gives project year discount factors.

Benefit/Cost Ratio.   The benefit/cost ratio (BCR) is used to compare alternatives whose costs may or may not be equal, whose benefits are unequal, and

Table 11-2. A comparison of "when to use" economic analysis techniques

TECHNIQUE	COSTS	BENEFITS	ECONOMIC LIFE	STATUS QUO
Present Value (PV) Analysis Discount future costs and benefits to present worth. Compares actual and discounted costs over project life	Necessary in any economic analysis			
Benefit/Cost Ratio (BCR) Indicates amount of benefit obtained per unit of cost	U/E	U	E	Not necessary
Uniform Annual Cost (UAC) Reduce alternatives' costs to streams of uniform annual costs	U	E	U	Not necessary
Savings/Investment Ratio (SIR) Relation between future savings and the investment needed to effect those	U/E	E	E	Necessary
Discounted Payback Analysis (DPA) Time period required for accumulated *PV* savings to offset *PV* investment costs	U/E	E	E	Necessary
Break-Even Analysis Point at which costs of alternatives are equal	U/E	E	E	Not necessary

U = unequal; E = equal.
Source: USDOT 1985.

Table 11-3. Project year discount factors

	TABLE A	TABLE B
	Present value of $1 (single amount— used when cash flows accrue in *varying* amounts each year)	Present value of $1 (cumulative uniform series—to be used when cash flows accrue in the *same* amount each year)
Project Year	10%	10%
1	0.954	0.954
2	0.867	1.821
3	0.788	2.609
4	0.717	3.326
5	0.652	3.978
6	0.592	4.570
7	0.538	5.108
8	0.489	5.597
9	0.445	6.042

Note: Table 11-3B factors represent the cumulative sum of Table 11-3A factors through any given project year.

whose economic lives are equal. The technique involves determining the life-cycle costs and benefits for each alternative and comparing them in ratio form. The alternative with the highest BCR is the preferred alternative.

The BCR is used for comparing alternatives for which the assumption of equivalent benefits is weak. The method of computing the BCR varies from analysis to analysis, depending upon whether or not the benefits are quantifiable.

The BCR is an economic indicator of cost effectiveness, computed by dividing benefits by costs for each alternative.

Many projects have a stated goal defined in terms of required output (e.g., to reduce errors, to decrease response time, to process an increased workload, etc.). The goal is not always quantified, but it is often susceptible to quantification and thus provides a potential measure of the benefits associated with the project.

Following are four examples of quantifiable output measures: (1) number of pages printed per hour, (2) number of reports generated per week, (3) number of work orders processed per month, and (4) decreased error rate per job.

The BCR technique can be used even when precise quantification of benefits is impossible. Using the ordinal ranking techniques, the analyst may assign an aggregate benefit value to each alternative. Because this method of evaluating benefits is subjective, the rationale for deriving the aggregate benefit values must be included in documenting the economic analysis.

Assumptions

1. The alternatives evaluated have unequal benefits.
2. The costs may or may not be equal.
3. The alternatives' economic lives are equal.

The method to determine the BCR of an alternative is as follows:

1. Choose a method for measuring the benefits of each alternative. If the benefits are readily quantifiable, determine them. If not, use the ordinal ranking technique.
2. Compute the PV costs of the alternatives.
3. Determine the BCR by dividing the quantified or weighted benefits by the PV costs.

No significance should be attached to the fact that a computed BCR may be less than 1. This is due entirely to the relative nature of the BCR and the arbitrarily chosen baseline. The only valid comparison is *between* the ratio measures.

Example

First, compute the PV cost for all fiscal years of each alternative. We know these costs from the PV analysis to be:

$$\text{Time sharing–status quo} = \$303,475$$
$$\text{Buy} = \$280,044$$
$$\text{Rent (six year lease)} = \$299,992$$

Next, determine the quantifiable benefit measures for each alternative, if possible. For instance, suppose the time-sharing alternative–status quo processes information correctly 95% of the time; the buy alternative processes information correctly 90% of the time and the rent alternative also processes information correctly 90% of the time. Establishing 1,000,000 pieces of information processed as an arbitrary amount, we can establish quantifiable benefit measures. For the first time-sharing option, it would be 950,000 pieces (95% of 1,000,000 pieces); for the buy option it would be 900,000 pieces (90% of 1,000,000); and for the rent option it would also be 900,000 pieces.

Thus, to calculate the BCR of each alternative, determine the ratio of benefits to costs.

$$\text{Time sharing} = 950,000/\$303,475 = 3.13$$

$$\text{Buy} = 900,000/\$280,044 = 3.21$$

$$\text{Rent} = 900,000/\$299,992 = 3.00$$

The buy option has the highest BCR.

Uniform Annual Cost. The uniform annual cost (UAC) technique is a cost-oriented approach to evaluating alternatives with unequal costs, equal benefits, and unequal economic lives. The technique involves stating all life-cycle costs for each alternative in terms of an average annual expenditure. The alternative with the lowest UAC is the most economical choice. It is not unusual for service lives to differ from alternative to alternative. When this occurs, the UAC can be used to put all the alternatives on a common basis of time to make a valid comparison.

Each alternative is converted into an equivalent hypothetical alternative having uniform recurring costs. The conversion is such that the total net PV costs of the actual alternative and its hypothetical equivalent are the *same*. The hypothetical alternatives are compared to determine the one with the lowest uniform recurring cost. The UAC is a hypothetical stream of annual or monthly payments made throughout the economic life of an alternative. The discounted value of these payments equals the actual PV cost of the alternative.

Assumptions

No end is foreseen to the requirement, nor do technological considerations play any significant role. It is, therefore, the limitation of *physical life* that constrains the economic lives of alternatives. The alternatives provide an equivalent level of benefits per year. Thus, even if these benefits are difficult to quantify, it is clear, in view of the unequal economic lives, that the *total* benefits afforded by the alternatives are not the same. The annual costs of the alternatives are unequal.

The following calculations can be used to obtain a UAC for each alternative:

1. Compute the PV cost for each alternative.
2. Divide the PV cost for all fiscal years by the sum of the discount factors for the economic life of the alternative (b_n). Cumulative discount factors are shown in Table 11-3B.

The UAC represents the amount of money which, if budgeted in equal yearly installments, would pay for the project. Note that this is not the same as taking a simple average. An example is a theoretical system with a five-year life and an acquisition cost of $2 million (m). If the UAC technique were used, the annual cost would be about $2.5m for the same building.

Simple Average	UAC
$\dfrac{\$10M}{5} = \$2M$	$\dfrac{PV}{b_n} = \dfrac{\$10M}{3.977} = \$2.5M$

The use of a simple average for determining the average annual cost for economic analysis is inappropriate because it fails to acknowledge the time value of money. The UAC, on the other hand, does incorporate this concept in its formula. In the above example, the significance of the $2.5M UAC is this:

If $2.5M were spent each year for five years, the total net PV of the payments would be $10M, the same as the actual net PV cost of the alternative.

Example

Suppose that in the previous example a lease had been signed to rent the system for eight years rather than six years. Because we now have an example of unequal project lives, the UAC of each alternative should be determined.

First, however, we should calculate the PV costs of year 7 and year 8 of the extended lease for the rent option. The PV costs are $32,998 for year 7 and $29,998 for year 8. These additional years result in a PV cost of an eight-year rent alternative of $362,989. (The rent option for years 1–6 = $299,992; $299,992 + $32,998 + $29,998 = $362,989.) (The sum of the parts does not equal the total due to rounding.)

Now the PV costs and economic lives of each alternative are as follows:

Status quo–time sharing	$303,475	6 years
Buy	$280,044	6 years
Rent	$362,989	8 years

To determine the UAC, look up the cumulative discount factor in Table 11-3B corresponding to the economic life of each alternative. The table shows that the cumulative discount factor is 4.570 for year 6 and 5.597 for year 8.

Thus, the UAC for each alternative is:

$$\text{Status quo–time sharing} = \frac{\$303,475}{4.570} = \$66,407$$

$$\text{Buy} = \frac{\$280,044}{4.570} = \$61,279$$

$$\text{Rent} = \frac{\$362,989}{5.597} = \$64,853$$

The buy alternative remains the least costly alternative.

The UAC analysis does not reflect actual cash outlays, but is used only for comparison as part of the decision-making process.

Savings/Investment Ratio.   The savings/investment ratio (SIR) technique is used to compare a proposed alternative to the status quo. The SIR is the ratio of discounted future cost savings to the discounted investment cost necessary to produce those savings. The alternative and the status quo may or may not have equal costs, but their benefits and economic lives are equal. The technique involves calculating three values: (1) the alternative's investment costs; (2) the life-cycle cost savings between the alternative and the status quo; and (3) the ratio of (2) to (1). The SIR determines the financial benefit attained from that alternative. For an investment to be economically sound, the SIR must exceed 1.

In an economic analysis, the SIR establishes a relationship between a proposed alternative and its status quo. When there is more than one alternative, the SIR technique will determine which alternative produces the most savings per dollar invested; however, it will *not* necessarily determine the least costly alternative. Consequently, the results of the SIR technique can be misleading, and it is suggested that this technique be reserved for analyses that compare a proposed alternative to the status quo.

When computing the SIR, the analyst is not interested in *total* operations costs—only in the difference between life-cycle operating costs for the two alternatives; that is the effect the investment has on the operation. Thus, the crucial question is, are the recurring savings of B relative to A sufficient to warrant the investment cost that would be necessary to implement alternative B? *Savings* means the reduced amount of annual expenditure resulting from replacement of the *status quo* by the proposed alternative.

The SIR is:

$$\text{SIR} = \frac{\text{PV(S)}}{I}$$

Because nonrecurring costs may be considered investment costs, they are used to determine the SIR. SIRs are useful because they provide the decision maker with a means of comparing the profitability of various investment projects. The SIR reflects the savings that will result for each dollar invested. For example, an investment with an SIR of 1.25 is more profitable than an investment with an SIR of 1.10 because it yields 15 cents more for each dollar invested.

Assumptions

1. The benefits of each alternative considered are equal.
2. The economic lives of each of the alternatives are equal.
3. There must be a status quo against which the alternatives are measured.

The method of calculating the SIR for each alternative, including the status quo, follows.

1. Compute the PV costs for each alternative.
2. Identify the nonrecurring costs and the month in which they occur for the alternatives being compared.
3. Determine the total PV of nonrecurring costs for all fiscal years for the alternatives. Write this amount on line (1) of the SIR format in Table 11-4. To perform this calculation:
   a. The subtotal line for nonrecurring costs × the discount factor for that month = the PV nonrecurring cost for that month.
   b. Sum the total PV nonrecurring costs and write the amount for that year in the right-hand box. You will probably have nonrecurring costs only in the first year. If necessary, calculate the PV nonrecurring cost for all fiscal years.
4. Subtract the total PV nonrecurring costs (see line 1) from the total PV cost for all fiscal years (line 2). This gives the total PV recurring costs for all fiscal years. Write this amount on line (3).
5. Determine the difference between the total PV recurring costs of the alternative and the total PV recurring costs of the status quo: line (4). This gives the PV savings of one alternative over the status quo and should be written on line (5).
6. Calculate the ratio of the PV savings, line (5), to the PV nonrecurring costs, line (1), for each alternative.
7. The higher the SIR, the greater the savings from each dollar invested.

The SIR is an economic analysis technique used to compare an alternative to the status quo when a significant investment is contemplated. The ratio calculated by relating one alternative to the status quo may then be compared to the ratio calculated by relating another alternative to the status quo. These

Table 11-4. SIR format

SIR ALTERNATIVE		BUY	RENT
(1) Total PV nonrecurring cost	= _____	$109,120	$ 19,840
(2) Total PV cost for all fiscal years	= _____	$280,044	$299,992
(3) Total PV recurring costs for all fiscal years: (2) − (1)	= _____	$170,924	$280,152
(4) Total PV recurring costs for status quo	= _____	$303,475	$303,475
(5) PV savings: (4) − (3)	= _____	$132,551	$ 23,323
PV savings			
SIR = PV nonrecurring costs = line (1)	= _____	1.21	1.18

ratios reflect the amount of savings returns for each dollar invested. The higher the SIR, the more financially attractive the alternative.

Discounted Payback Analysis.   The discounted payback analysis (DPA) technique is used to compare a proposed alternative to the status quo. The alternative and the status quo may or may not have equal costs, but their benefits are equal, as are their economic lives. The technique involves calculating:

1. the alternative's investment costs.
2. the life-cycle cost savings between the alternative and the status quo.
3. the ratio of (1) to (2).

This ratio, which is simply the inverse of the SIR, determines the period required for a project's accumulated savings to offset the investment costs. Each alternative may be compared to the status quo; the alternative with the lowest ratio has the quickest payback period. Often, project reviewers are interested in knowing when a project will pay for itself. When that question arises, it is necessary to calculate the payback period. DPA calculates the payback period, the time elapsed between the initial investment and the point at which the payback on that investment will occur. The calculation of a payback period is not affected by the duration of the project's economic life. For example, a 4.5-year payback period means the same thing whether the economic life is 10 years or 25 years.

The method you should use to calculate the discounted payback period for each alternative, including the status quo, is as follows:

1. Compute the PV costs for each alternative.
2. Identify the nonrecurring costs and the month in which they occur for the alternative.
3. Determine the total PV of the nonrecurring costs for each fiscal year of the alternative. Then multiply the nonrecurring costs by the discount factor for each month to obtain the PV nonrecurring costs for that month.
4. Sum the total PV nonrecurring costs. You will probably have nonrecurring costs only in the first year, but if necessary, calculate the PV nonrecurring costs for all fiscal years.
5. Subtract the total PV nonrecurring costs from the total PV for the fiscal year of the alternative to determine the total PV recurring cost for each fiscal year, and subtract this from the total PV recurring cost for the fiscal year of the status quo.
6. Repeat the entire process for each fiscal year of the project.
7. Then determine the cumulative savings for each year. Subtract the cumulative nonrecurring costs from the cumulative savings to determine the cumulative cash flow. Then divide this by the savings for the following

fiscal year and multiply by 12 (months). This is the month during which the discounted payback occurs for the first year with a positive cumulative cash flow.

With DPA the payback is biased toward alternatives having low investment costs, since they are paid back quickly, and there is no means to compare alternatives other than buy (lease, rent). DPA can be used to prioritize alternatives but should not be used alone for selection.

Break-Even Analysis. Break-even analysis (BEA) is a technique used to display graphically the relation between alternative cost patterns. It may be used to compare alternatives whose costs may or may not be equal, whose benefits are equal, and whose economic lives are equal. The technique involves finding the point at which the costs are equivalent. To either side of the break-even point, one alternative or the other has the economic advantage. When an alternative is compared against the status quo (if one exists), the break-even point determines when savings will begin to accrue.

BEA is also a useful tool for analyzing the financial characteristics of one or more alternatives when relative desirability depends on the quantity of some variable, such as the number of units produced or the number of hours of system operation.

Assumptions

1. More than one alternative is being considered.
2. The nonmonetary benefits of the alternatives compared are equal or are irrelevant for purposes of this comparison.
3. The economic lives of the compared alternatives are equal.
4. Variable cost components may be involved, such as hours of computer time or units produced.

Calculation

1. Determine the total PV cost for the first month of the procurement. Add that to the total PV cost for the second month, and repeat this process for each month of the project's or system's life.
2. For carryover between fiscal years, the last entry in the cumulative PV row for each fiscal year should be added to the PV cost of the first month of the next fiscal year.
3. Display the PV cost lines on an appropriately labeled graph. The vertical axis should be labeled in units of cumulative cost for each month of the alternative and the horizontal axis in units of time. The intersection of the lines, if any, should be identified as the break-even point.

Example

To determine the break-even point of our example, we simply graph the cumulative PV costs over all fiscal years and determine where, if at all, the lines intersect. At that point of intersection, the cumulative PV costs are equal and the decision maker is indifferent to the costs of the alternatives (Fig. 11-1).

In this example, the break-even point occurs during May, fiscal year 1992, for the buy and rent options; during June, fiscal year 1991, for the buy and status quo options; and during August, fiscal year 1992, for the rent and status quo options.

Sensitivity Analysis

Because of uncertainties in the analysis, the decision maker will want to know not only the economic choice implied by the best estimate of costs and benefits, but also whether that decision would change if one or more of the inputs varied. Sensitivity analysis, described here (USDOT 1985), provides this information.

Sensitivity refers to the relative magnitude of change in one or more elements of an economic analysis that will cause a change in the ranking of alternatives. In a sensitivity analysis, if one particular factor or cost element can be varied over a wide range without affecting the ranking of alternatives, the analysis is said to be *insensitive* to uncertainties regarding that particular element.

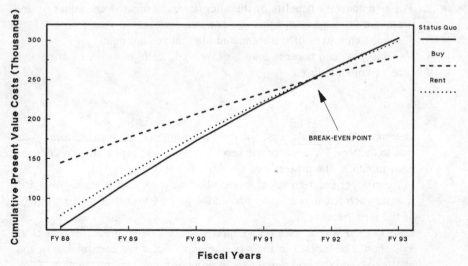

Figure 11-1. Break–even analysis result.

Sensitivity analysis does not require sophisticated techniques, but rather the ability to recognize uncertainties in the economic analysis and to deal with them logically. The analysis may be reworked to test sensitivity to such elements as cost estimates, length of system life, volume or mixture of work, equipment configuration, requirements, and assumptions.

First, the analyst must determine whether a sensitivity analysis is necessary. If there is complete certainty and the ranking of alternatives establishes one option as markedly superior to the rest, the analyst should not be concerned about testing for sensitivity. It is only when there is uncertainty and the economic choice is *not* clear that further investigation is required.

If a sensitivity analysis is indicated, the analyst must select parameters to test. Sensitivity analysis should treat dominant input variables, or those having a significant impact on the total PV cost or the benefits. The choice of input variables for sensitivity may depend not only on relative dominance, but also on the degree of confidence that can be placed in these estimates.

The basic procedure for sensitivity testing has four steps: first, select the factor to be tested; second, hold all parameters in the analysis constant except that factor; third, rework the analysis, using different estimates for the factor under consideration; and fourth, check the results. If the ranking of alternatives is affected, the analysis is sensitive to that amount of change in that variable.

Each key parameter should be tested individually to determine its effects on the analysis.

Example

1. Given the following cost data, determine the less costly alternative:

	ALTERNATIVE A PROPOSED	ALTERNATIVE B STATUS QUO
Year 1:		
ADPE	$ 80	$ 80
System development	180	0
Site preparation	35	0
Years 2–9		
Personnel	$ 80/yr	$120/yr
Other operating		
costs	20/yr	25/yr

Note: Dollars in thousands.

2. Will the results change if the system development costs are $200,000? $210,000?
3. What will be the impact if personnel costs are increased to $85,000 per year?

Solution

1. The net PV value for alternatives A and B are as follows:

$$PV \text{ (alt. A)} = 0.954(\$80 + \$180 + \$35) + 5.088(\$80 + \$20)$$
$$= \$281 + \$509$$
$$= \$790,000$$
$$PV \text{ (alt. B)} = 0.954(\$80) + 5.088(\$120 + \$25)$$
$$= \$76 + \$738$$
$$= \$814,000$$

Thus, alternative A, the proposed system, is the less costly.

2. If the system development cost is $200,000:

$$PV \text{ (alt. A)} = 0.954(\$80 + \$200 + \$35) + 5.088(\$80 + \$20)$$
$$= \$301 + \$509$$
$$= \$810,000$$

Since $810,000 is less than the cost of the status quo alternative, the analysis is not sensitive to a $20,000 increase in system development costs.

If the system development cost is $210,000:

$$PV \text{ (alt. A)} = 0.954(\$80 + \$210 + \$35) + 5.088(\$80 + \$20)$$
$$= \$310 + \$509$$
$$= \$819,000$$

In this example, the cost would be greater for the proposed system. Therefore, the analysis is sensitive to a $30,000 increase in system development.

3. If annual personnel costs are increased by $5,000, then:

$$PV \text{ (alt. A)} = 0.954(\$80 + \$180 + \$35) + 5.088(\$85 + \$20)$$
$$= \$281 + \$534$$
$$= \$815,000$$

Thus, the analysis *is* sensitive to this change.

Example

The economic life in the above example is somewhat questionable. Perform a sensitivity analysis to determine what would happen if the economic life was six years instead of nine. Based on a six-year economic life, the PV values of alternatives A and B are as follows:

$$PV \text{ (alt. A)} = 0.954(\$295) + 3.616(\$100) = \$643,000$$

$$PV \text{ (alt. B)} = 0.954(\$80) + 3.616(\$145) = \$600,000$$

Alternative B is now less costly than Alternative A. Since the ranking of alternatives has changed, the analysis *is* sensitive to the shorter economic life.

DECISION TREES

The decision tree is a technique for lining out all possible outcomes for each alternative to be compared and then assigning probabilities to each of these outcomes. The sum of the probabilities for each alternative must equal 1. Each assigned probability is multiplied by the presumed payoff to give an expected value. The expected values within each alternative are summed; then they are compared to indicate the most desirable alternative or the choice most likely to maximize benefits. The process is useful in helping the decision maker think about and consider the outcomes.

The decision tree concept has been developed and refined. There are now a variety of decision support systems that use powerful computer capabilities to handle complicated and voluminous information. The analytic hierarchy process uses the decision tree concept in a detailed and sophisticated approach that includes the handling of qualitative information. Decision trees can also be used to help develop alternatives for economic analyses.

REFERENCES AND RECOMMENDED READINGS

Acroff, R. L. 1970. A Concept of Corporate Planning. New York: Wiley-Interscience.

Arrow, K. J. and R. C. Lind. 1970. Uncertainty and the evaluation of public investment decisions. Am. Econ. Rev. 60:364–378.

Bentkover, C. M. 1986. Benefits Assessment: State of the Art. Dordrecht, The Netherlands: D. Reidell Publishing Company.

Campen, J. T. 1986. Benefit, Cost, and Beyond: The Political Economy of Benefit Cost Analysis. Cambridge, MA: Ballinger Publishing Company.

Canada, J. R. 1971. Intermediate Economic Analysis for Management and Engineering. Englewood Cliffs, NJ: Prentice-Hall, Inc.

Clark, R. D. and R. T. Lackey. 1976. A technique for improving decision analysis in fisheries and wildlife management. VA J. Sci. 27:199–201.

Clark, T. E. 1974. Decision-making in technologically based organizations: A literature survey of present practice. IEEE Trans. Eng. Manag. EM-21(1):9–23.

Crowder, L. B. and J. J. Magnuson. 1983. Cost-benefit analysis of temperature and food resource use: A synthesis with examples from the fishes. *In* Behavioral Energetics, ed. W. P. Aspey and S. I. Lustick, pp. 189–221. Columbus: Ohio State University Press.

Dalkey, N. and O. Helmer. 1963. An experimental application of the Delphi method to the use of experts. Management Sci. 9:458–467.

Daubert, J. T. and R. A. Young. 1981. Recreational demands for maintaining instream flows: A contingent valuation approach. Am. J. Agric. Econ. 63:666-676.

Dixon, J. A. and M. M. Hufschmidt. 1986. Economic Valuation Techniques for the Environment. Baltimore, MD: Johns Hopkins University Press.

Fusfeld, A. R. and R. N. Foster. 1971. The Delphi technique: Survey and comment. Bus. Horizons 14:63–74.

Hirshleifer, J. 1966. Investment decisions under uncertainty: Applications of the state preference approach. Q. J. Econ. 80:252–277.

Levin, H. M. 1983. Cost-Effectiveness, A Primer, New Perspectives in Evaluation, Volume 4. Beverly Hills, CA: Sage Publications.

Linstone, H. A. and M. Turoff. 1975. The Delphi Method. Reading, MA: Addison-Wesley Publishing Company.

MacDonald, C. R., T. L. Meade, and J. M. Gates. 1975. A production cost analysis of closed system culture of salmonids. Marine Tech. Rep. No. 41. Narragansett, RI: University of Rhode Island.

Magee, J. F. 1964a. Decision trees for decision making. Harvard Bus. Rev. 42:126–138.

Magee, J. F. 1964b. How to use decision trees in capital investment. Harvard Bus. Rev. 42:79–93.

Martino, J. P. 1972. Technological Forecasting for Decision Making. New York: American Elsevier Publishing Company.

McNeil, W. J. and J. E. Bailey. 1975. Salmon Rancher's Manual. Auke Bay, AK: National Oceanic and Atmospheric Administration, Northwest Fisheries Center, Auke Bay Fisheries Laboratory.

United States Dept. of Commerce. 1980. Life-Cycle Cost Manual for the Federal Energy Management Program. NBS Handbook 135. Washington, DC.

United States Department of Energy. 1982. Cost Guide, Volume I, Economic Analysis: Methods, Procedures, Life Cycle Costing, and Cost Review/Validating. DOE/MA-0063. Washington, DC: DOE.

United States Department of the Treasury (USDOT). 1985. Economic analysis guidelines for ADP resources. IRS Document 7069 (12-85). Washington, DC: USDOT.

United States Office of Management and Budget. 1972. Discount Rates to be Used in Evaluating Time Distributed Costs and Benefits. OMB Circular No. A-94, revised. Washington, DC.

Zurboy, J. R. 1981. A new tool for fishery managers: The Delphi technique. N. Am. J. Fish. Mgmt. 1:55–59.

Chapter 12

Computer Assisted
Decision Support Systems

FOCUS: Methodology is now available to analyze the most complex problems and produce optimum (maximum or minimum) solutions.

HIGHLIGHTS:
- Quantifying the qualitative factors
- Selecting an oxygenation system
- Maximizing hatchery effectiveness
- Maximizing returns of Atlantic salmon

ORGANIZATION: Analytic Hierarchy Process
  AHP Decision Analysis for Procurement,
    Installation, and Use of Oxygenation Equipment
Linear Programming

ANALYTIC HIERARCHY PROCESS

The analytic hierarchy process (AHP) provides a logical, organized means to handle all important qualitative as well as quantitative factors in the decision-making process. The AHP and similar support systems require the use of a computer and appropriate software. AHP uses some fancy mathematical footwork, the eigenvalue approach, to enable the valid use of quantitative judgments in real number operations (by converting all measurements to ratios). The theory makes interesting reading, even to some nonmathematicians. But it is not necessary to understand the theoretical derivation of AHP in order to use it.

As in other decision support systems, the first step is to define the problem or objective. Next, one must identify the important criteria that affect the solution to the problem or the means to reach the objective—the things that affect the desirability of each alternative. These include the important uncertainties that are varied in a sensitivity analysis. For each set of criteria there is a set of alternatives or subcriteria: each of these may have a set of alternatives, and so on. Beginning with the first set of criteria, each criterion is compared, one at a time, with each of the others. Judgments are made about their relative, comparative importance in this pairwise fashion. Priorities for criteria are then based on these pairwise judgments. An example will clarify the process.

AHP Decision Analysis for Procurement, Installation, and Use of Oxygenation Equipment

Suppose the manager of a fish hatchery has identified three alternatives for a system to control dissolved gases: (1) a liquid oxygen system (LOX), (2) a pressure swing adsorption (PSA) oxygen generator, and (3) no change (status quo). When the manager lists the pros and cons of each alternative, it is evident that there are many important considerations in addition to the results of a cost-benefit analysis. There are questions regarding the system's reliability and maintenance; its ability to alleviate nitrogen gas problems and low flow situations; important but possibly embarrassing concerns about political prestige, employee morale, and the manager's vanity; and several financial considerations in addition to the PV calculations (cost-benefit determination), such as payback period and control (as the ability to reevaluate and alter or cancel the system, if appropriate).

The problem quickly becomes complex and confusing. We organize the objectives as criteria and subcriteria in some manner such as the following:

Financial considerations
   Present value
   Discounted cash payback period
   Control
Staffing considerations
   Employee morale
   Training
   Flexibility for future needs
Political considerations
   Labor changes (ceilings)
   Manager's (individual's) vanity
   Prestige in the public eye (press)
Serviceability
   Low flow usefulness
   Nitrogen gas dissipation
   Power outage and requirement
   Control over dissolved oxygen

Figure 12-1 shows the criteria and subcriteria arranged for AHP, and Figure 12-2 illustrates the arrangement with the addition of the three alternatives.

Priorities are determined from pairwise comparisons, made by either a verbal or a numerical method. To use a verbal mode, choose one of five levels

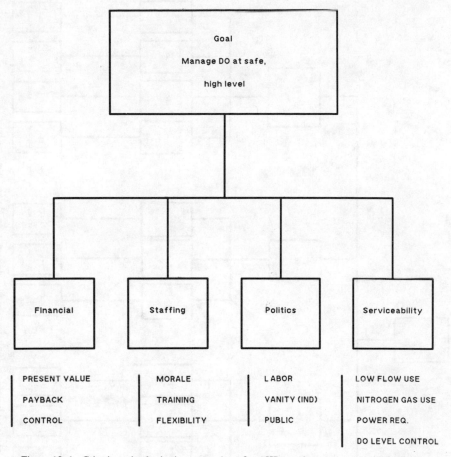

Figure 12–1. Criteria and subcriteria arrangement for AHP to select an oxygenation system.

of importance, or choose between any two consecutive levels. For instance, "Alternative 1 is (*your choice*) more important than alternative 2." This is done for each possible pair (six times), using the following choices:

Extremely
Very strongly
Strongly
Moderately
Equally

In this example, the manager judges (chooses) politics to be moderately more important than finance, which is judged to be much (strongly) more

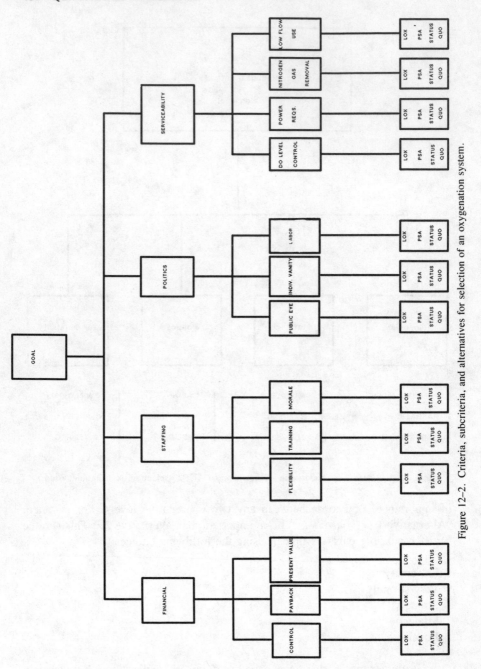

Figure 12–2. Criteria, subcriteria, and alternatives for selection of an oxygenation system.

	Present Value	Payback	Control
PRESENT VALUE		5.0	(3.0)
PAYBACK			(6.0)
CONTROL			

Matrix entry indicates that ROW element is:

1 EQUALLY, 3 MODERATELY, 5 STRONGLY, 7 VERY STRONGLY, 9 EXTREMELY

more IMPORTANT than COLUMN element

unless enclosed in parentheses.

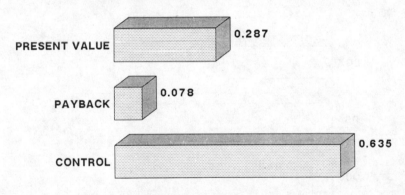

PRESENT VALUE   0.287

PAYBACK   0.078

CONTROL   0.635

INCONSISTENCY RATIO = 0.081

Figure 12–3. Judgments and priorities with respect to the goal of selecting the best alternative for oxygenation.

important than serviceability. Figure 12-3 indicates the results of prioritizing the criteria. Other comparisons in Figure 12-3 show that politics is moderately more important than staffing, which is moderately to strongly more important than serviceability. Politics is very strongly more important than serviceability. Relative priorities were calculated on these judgments and are shown decimally and graphically. An inconsistency ratio is used to indicate mistakes or whether you have deceived yourself in making one or more comparisons. An inconsistency ratio exceeding 0.10 indicates that a review of the comparisons (your judgments) is needed.

Judgments and the calculated priorities for the financial considerations node are summarized in Figure 12-4. The 0.287 for PV, 0.078 for payback,

	LOX	PSA	STAT QUO
LOX		(3.0)	7.0
PSA			9.0
STAT QUO			

Matrix entry indicates that ROW element is:

1 EQUALLY, 3 MODERATELY, 5 STRONGLY, 7 VERY STRONGLY, 9 EXTREMELY

more IMPORTANT than COLUMN element

unless enclosed in parentheses.

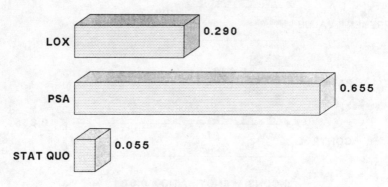

INCONSISTENCY RATIO = 0.069

Figure 12–4. Judgments and priorities with respect to the financial subcriteria.

and 0.635 for control reflect a judgment that PV is strongly more important than payback; control is moderately more important than PV; and control is strongly to very strongly more important than payback.

Judgments about each subcriterion are made similarly. Figure 12-5 shows relative preferences for the alternatives regarding the public eye (press) aspects of politics. If the manager were concerned only with the public relations aspect of oxygenation, the PSA would be by far the system of choice.

Figures 12-6 and 12-7 present the priorities for criteria, subcriteria, and alternatives as "local" and "global" models. The global priority of the PUBLIC node (0.054) is the product of its local priority (0.105) and its parent node (POLITICS) global priority (0.515). The priorities of the subcriteria are additive, and together equal the priority of the parent criteria.

	Financial	Staffing	Politics	Serviceability
Financial		(3.0)	(3.0)	5.0
Staffing			(3.0)	4.0
Politics				7.0
Serviceability				

Matrix entry indicates that ROW element is:

1 EQUALLY, 3 MODERATELY, 5 STRONGLY, 7 VERY STRONGLY, 9 EXTREMELY

more IMPORTANT than COLUMN element

unless enclosed in parentheses.

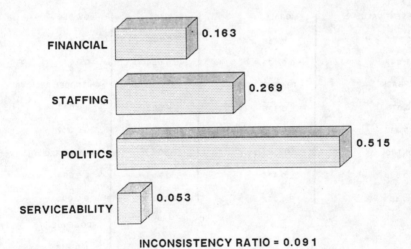

FINANCIAL   0.163

STAFFING   0.269

POLITICS   0.515

SERVICEABILITY   0.053

INCONSISTENCY RATIO = 0.091

Figure 12–5. Judgments and priorities for alternatives with respect to the public view subcriteria (of the politics criteria).

All the judgments have been synthesized, and the alternative preferences sorted and presented in Table 12-1. The preferences (high to low) are arranged from top to bottom. Figure 12-8 shows the results. The liquid oxygen system (LOX) is the most preferable and the status quo the least preferable. The difference, or preference for the LOX system, in this example is substantial.

A sensitivity analysis, however, would indicate circumstances under which the decision would change. Figure 12-9 illustrates the sensitivity analysis of results on the relative importance of the POLITICS criteria. The horizontal axis is the priority of the POLITICS criteria, and the vertical axis represents alternative priorities. The dashed vertical line indicates the relative priorities from judgment in Figure 12-8; it is the location of AHP results on this graph,

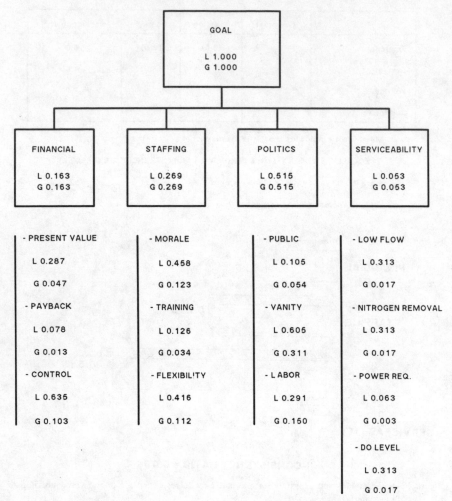

Figure 12–6. Local and global priorities viewed from the global node.

and shows that POLITICS has a roughly 52% local priority. If the priority of POLITICS were decreased, the dashed vertical line would move to the left. Its intersection with LOX would indicate a progressively lower value for LOX, whereas it would intersect with status quo at progressively higher values.

LINEAR PROGRAMMING

Linear programming (LP) is a field of study and applications in mathematics. *Linear* means that the functions used are simple linear functions; none are raised to a power other than 1. *Programming* means "getting from here to there" in

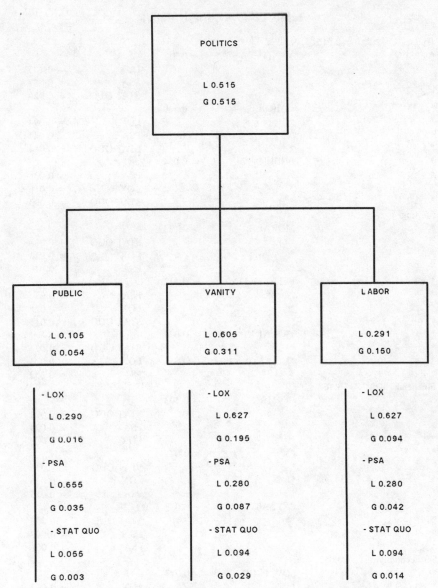

Figure 12–7. Local and global priorities viewed from the politics node.

Table 12-1. Synthesis details

Politics	= 0.515				
*		VANITY	= 0.311		
*		*		LOX	= 0.195
*		*		PSA	= 0.087
*		*		STAT QUO	= 0.029
*		LABOR	= 0.150		
*		*		LOX	= 0.094
*		*		PSA	= 0.042
*		*		STAT QUO	= 0.014
*		PUBLIC	= 0.054		
*		*		PSA	= 0.035
*		*		LOX	= 0.016
*		*		STAT QUO	= 0.003
Staffing	= 0.269				
*		MORALE	= 0.123		
*		*		STAT QUO	= 0.088
*		*		LOX	= 0.018
*		*		PSA	= 0.018
*		FLEXIBILITY	= 0.112		
*		*		PSA	= 0.075
*		*		LOX	= 0.025
*		*		STAT QUO	= 0.011
*		TRAINING	= 0.034		
*		*		STAT QUO	= 0.027
*		*		LOX	= 0.004
*		*		PSA	= 0.002
Financial	= 0.163				
*		CONTROL	= 0.103		
*		*		STAT QUO	= 0.072
*		*		PSA	= 0.024
*		*		LOX	= 0.008
*		PV	= 0.047		
*		*		STAT QUO	= 0.016
*		*		LOX	= 0.016
*		*		PSA	= 0.015
*		PAYBACK	= 0.013		
*		*		STAT QUO	= 0.009
*		*		LOX	= 0.002
*		*		PSA	= 0.001
Serviceability	= 0.053				
*		LOW FLOW	= 0.017		
*		*		LOX	= 0.012
*		*		PSA	= 0.003
*		*		STAT QUO	= 0.001
*		N_2 REMOVAL	= 0.017		
*		*		STAT QUO	= 0.012
*		*		PSA	= 0.003
*		*		LOX	.97E-03

Table 12-1 (cont.)

*	DO LEVEL		= 0.017
*	*	LOX	= 0.008
*	*	PSA	= 0.008
*	*	STAT QUO	= 0.001
*	POWER REQ.		= 0.003
*	*	STAT QUO	= 0.003
*	*	PSA	.59E-03
*	*	LOX	.24E-03

an orderly, sensible fashion. For those unfamiliar with LP, skimming some LP references can provide an enjoyable and enlightening experience—the story of two separate, and simultaneously developed, approaches coming together to form an exciting basis for optimizing and a variety of other procedures that have come out of the field like popcorn out of a frying pan.

Computers have been used to wire the power of LP into a set of micro switches small enough to be thrown back and forth by the mental muscle of the average fish culturist. . . . and even by that of some college professors who don't teach in mathematics departments. Computer packages can be used for developing the marketing mix, maximizing returns to resources through enterprise mixing, minimizing transportation, logistics, or feed ingredients, and more. The example given here illustrates the value of linear programming to the public resource manager who was trying to restore Atlantic salmon to a river system in Chapter 2.

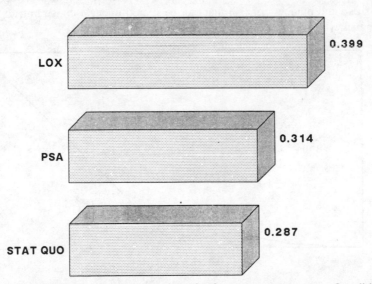

Figure 12–8. Synthesis results for the best alternative for an oxygenation system. Overall inconsistency index = 0.080.

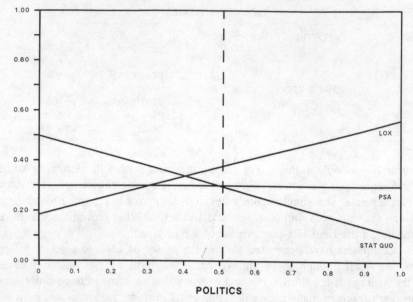

Figure 12–9. Sensitivity analysis.

First, we assume that larger fish survive at a higher rate than smaller fish, and from that develop or use a size-related survival curve. Figure 12-10 shows three possibilities. Second, we assume that the hatchery has a finite capacity (or a finite budget for food and labor) and that there must be a tradeoff between size of fish and numbers of fish that can be reared under a constraint; to grow bigger fish, you must grow fewer fish.

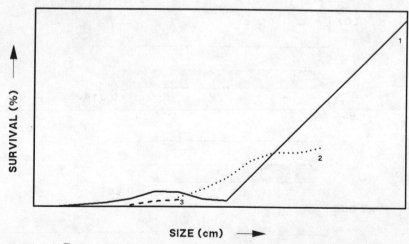

Figure 12–10. Three survival curves relating the size of yearling Atlantic salmon smolts (at release) to survival (returned from the ocean as adults).

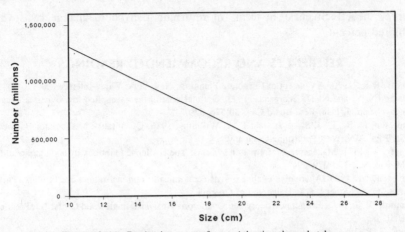

Figure 12-11. Production curve for an Atlantic salmon hatchery.

By using a computer software package for LP, one can enter a set of production numbers, such as points for the production line shown in Figure 12-11, and a set of points describing a survival relation (Figure 12-10), and maximize for the number of returning salmon as the "decision" variable. The results of this relatively simple procedure, using one production curve and identical constraints for each of three survival curves, are shown in Figure 12-12. They indicate that the production size approach, or the stocking size mix, should be geared to the estimated survival data. The three survival curves indicate that three different production scenarios are needed. If the production manager is simply using a scenario based on a similar but incorrect survival

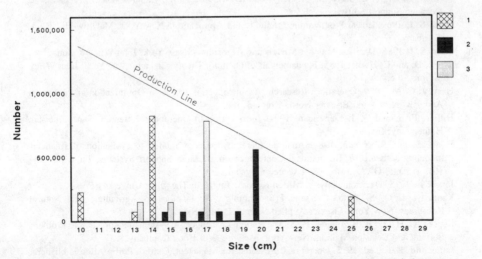

Figure 12-12. Production mixture of the number and size of Atlantic salmon that maximize returns based on three survival curves.

curve, the effectiveness in terms of returning fish could change by several hundred percent.

REFERENCES AND RECOMMENDED READINGS

Acroff, R. L. 1970. A Concept of Corporate Planning. New York: Wiley-Interscience.

Ahmed, N. V. and N. D. Georganas. 1973. Optimal control theory applied to a dynamic aquatic eco-system. J. Fish. Res. Board Can. 30:576–579.

Anderson, D. R., D. J. Sweeney, and T. A. Williams. 1986. Quantitative Methods for Business. St. Paul, MN: West Publishing Company.

Argyris, C. 1971. Management information systems: The challenge to rationality and emotionality. Manag. Sci. 17:B279–294.

Arraes, R. A. 1983. Alternative evaluations of economically optional rations for broilers. Ph.D. dissertation. Athens, GA: University of Georgia.

Baumol, W. J. 1977. Economic Theory and Operations Analysis. Englewood Cliffs, NJ: Prentice-Hall, Inc.

Brans, J. P. 1984. Operational Research '84: Proceedings of the Tenth International Conference on Operational Research. New York: Elsevier Science Publishing Company.

Clark, R. D., Jr. and R. A. Martin. 1986. Minimizing Cost of Transporting Fish from Hatcheries to Public Fishing Waters. Michigan DNR Fish. Res. Rep. No. 1939.

Cornell, A. H. 1980. The Decision Maker's Handbook. Englewood Cliffs, NJ: Prentice-Hall, Inc.

Dodge, J. A. 1981. Buy or lease cost model—selected railway equipment. Final Report. Fort Lee, VA: Logistics Studies Office.

Etlon, S. 1969. What is a decision? Manag. Sci. 16:B172–189.

Etzioni, A. 1967. Mixed scanning: A "third" approach to decision making. Public Admin. Rev. 27:385–392.

Forman, E. H. 1983. The analytic hierarchy process as a decision support system. Proc. IEEE Computer Soc. pp. 251–272.

Forman, E. H. 1985. Decision support for executive decision makers. Information Strategy: Executive J. 1(4).

Forman, E. H., T. L. Sasty, M. A. Selly, and R. Waldron. 1983. Expert Choice. McLean, VA: Decision Support Software.

Gass, S. I. 1969. Linear Programming Methods and Application. New York: McGraw-Hill Book Company.

Gass, S. I. 1985. Decision Making, Models and Algorithms. New York: John Wiley & Sons, Inc.

Gross, D. and C. Harris. 1985. Fundamentals of Queuing Theory, 2nd ed., New York: John Wiley & Sons, Inc.

Harvey, C. M. 1979. Operations Research: An Introduction to Linear Optimization and Decision Analysis. New York: Elsevier North–Holland, Inc.

Hillier, F. S. and G. J. Lieberman. 1974. Introduction to Operations Research. San Francisco: Holden-Day, Inc.

Hong, S. and R. Nigam. 1981. Analytic hierarchy process applied to evaluation of financial modeling software. In International Conference on Decision Support Systems, First. pp. 96–101. Atlanta, GA. Austin, TX: Execucom Systems.

Jones, L. 1975. Decision analysis. Milton Keynes, England: The Open University Press.

Kennedy, J. O. S. 1986. Dynamic Programming: Applications to Agriculture and Natural Resources. New York: Elsevier Applied Science Publishers.

Kepner, C. H. and B. B. Tregoe. 1965. The Rational Manager: A Systematic Approach to Problem Solving and Decision Making. New York: McGraw-Hill Book Company.

Littlechild, S. C., ed. 1977. Operational Research for Managers. Oxford: Philip Allen Publishers, Ltd.

Luce, R. D. and H. Raiffa. 1959. Games and Decisions, chap. 7. New York: John Wiley & Sons, Inc.

MacCrimmon, K. R. 1973. An overview of multiple objective decision making. *In* Multiple Criteria Decision Making, ed. J. L. Cochrane and M. Zeleny, pp. 18–44. Columbia, SC: University of South Carolina Press.

MacCrimmon, K. R. 1974. Managerial decision making. *In* Contemporary Management—Issues and Viewpoints, ed. J. W. McGuire, pp.445–495. New York: Prentice-Hall, Inc.

Mendelssohn, R. 1982. Discount factors and risk aversion in managing random fish populations. Can. J. Fish. Aqua. Sci. 39:1252–1257.

Miller, G. A. 1956. The magical number seven, plus or minus two: Some limits on our capacity for information processing. Psych. Rev. 63:81–97.

Moder, J. J. and C. R. Phillips. 1964. Project Management with C.P.M. and P.E.R.T. New York: Reinhold Publishing Company.

Raiffa, H. and R. Schlaifer. 1978. Applied Statistical Decision Theory. Cambridge, MA: M.I.T. Press.

Romero, C. and T. Rehman. 1984. Goal programming and multiple criteria decision-making in farm planning: An expository analysis. J. Agric. Econ. 35:177–190.

Saaty, T. L. 1980. The Analytic Hierarchy Process. New York: McGraw-Hill Book Company.

Saaty, T. L. 1982. Decision Making for Leaders. Belmont, CA: Lifetime Learning Publications, a division of Wadsworth Publishing Company, Inc.

Sale, M. J., E. D. Brill, Jr., and E. E. Herricks. 1982. An approach to optimizing reservoir operation for downstream aquatic resources. Water Resou. Res. 18:705–712.

Schaifer, R. 1969. Analysis of Decisions Under Uncertainty. New York: McGraw-Hill Book Company.

Simon, H. A. 1960. The New Science of Management Decision, pp. 40–43. New York: Harper & Brothers.

Singh, J. 1972. Great Ideas of Operations Research. New York: Dover Publications.

Stamp, N. H. E. 1978. Computer technology and farm management economics in shrimp farming. Proc. World Maricul. Soc. 9:383–392.

Tomlinson, J. W. C. and P. S. Brown. 1979. Decision analysis in fish hatchery management. Trans. Am. Fish. Soc. 108:121–129.

Wagner, H. M. 1975. Principles of Operations Research, pp. 667–672. Englewood Cliffs, NJ: Prentice-Hall, Inc.

Woolsey, R. E. D. and H. S. Swanson. 1975. Operations Research for Immediate Application: A Quick and Dirty Manual. New York: Harper & Row Publishers, Inc.

Wright, S. 1981. Contemporary Pacific salmon fisheries management. N. Am. J. Fish. Manag. 1:29–40.

Appendix I

Manager Attributes and Expectations

Desirable Attributes and Characteristics of Managers
(Suggested by Various Sources)
Distinguishing Attributes and Characteristics of Many
Successful Managers (Hammaker and Rader 1977)
U.S. Fish and Wildlife Service Expectations for an
Entry-Level Manager
Characterization of Immediately Promotable
Supervisors (Likert 1967)
Traits of Top Managers (Osburn and
Schneeberg 1983)

DESIRABLE ATTRIBUTES AND CHARACTERISTICS OF MANAGERS (SUGGESTED BY VARIOUS SOURCES)

Maintains a high level of productivity and cost effectiveness.
Maintains and inspires enthusiasm.
Develops confidence and cohesiveness among co-workers and subordinates.
Involves subordinates in planning and decision making.
Accepts responsibility for problems.
Listens actively and communicates openly.
Supports employees' development.
Delegates authority.

DISTINGUISHING ATTRIBUTES AND CHARACTERISTICS OF MANY SUCCESSFUL MANAGERS (Hammaker and Rader 1977)

Attributes
  Self-motivation
  Emotional maturity

147

Common sense and good judgment
Sensitivity to people
Inquiring minds
Average to superior intelligence
Integrity
Characteristics
Like people.
Can select a person who will perform well in a specific job.
Can select persons who have growth potential.
Are interested in and concerned about both people and production.
Use help from subordinates and superiors.
Earn the support and good will of subordinates.
Use good judgment, are fair and consistent, believe in people, help people grow, defend subordinates in need if it can be done in good conscience, and instill self-confidence by treating subordinates as individuals of dignity and worth.
Punish failure and reward success.
Know the business, its objectives, purposes, processes, and methods.
Become vigorous and ethical competitors.

U.S. FISH AND WILDLIFE SERVICE EXPECTATIONS FOR AN ENTRY-LEVEL MANAGER

Management expectations of the first-line supervisor:

Work
Work within the organization to accomplish the section's goals and contribute to the organization's goals.
Keep up-to-date technically in order to supervise and develop subordinates technically.
Keep employees constructively occupied.
Identify problems, deal with their causes, and work to improve the section.
Control the assigned work and review results for quality while operating economically.
Communication
Keep upper management informed about activities and problems.
Keep employees informed.
Human Relations
Think of oneself as a member of management.
Support management's problems.
Handle complaints from subordinates.

Employees' expectations of the first-line supervisor:

Work
Be technically competent, and able to supervise and develop employees technically.
Provide increasingly responsible and challenging work assignments.
Provide good working conditions—space, equipment, and materials.
Communication
Let employees know what management expects of them.
Fairly present employees' point of view to upper management.
Identify with employees and support them in dealings with upper management and the public.
Human Relations
Maintain good morale and good relations within the group.
Recognize employees publicly for good performance.
Criticize employees constructively, privately, and as soon after the employee makes a mistake as possible.
Treat all employees fairly and equally.

CHARACTERIZATION OF IMMEDIATELY PROMOTABLE SUPERVISORS (Likert, 1967)

Good at handling people; doesn't have a lot of problem employees; and seems to get along well with employees.
Employees feel free to discuss personal problems, as well as important aspects of the job, with this supervisor. Gets to know employees as individuals and helps them adjust.
Lets employees know where they stand. Of the employees under supervisors rated immediately promotable, 79% believed they had a good idea of what the supervisor thought of their work. Only 42% of those working for questionable supervisors felt they knew where they stood. Good supervisors evaluate employees informally as a normal part of work relations.
Stands up for employees, who are confident that the supervisor is both willing and able to represent them.
Tries to resolve employee complaints.
Lets employees work generally on their own, and does not give detailed and frequent instructions that limit freedom.
Holds frequent group discussions. The frequency of group meetings is a rough measure of the extent to which a supervisor communicates. Such meetings specify goals and give each person a feeling of responsibility for the success of the decision. In one organization, it was found that

the more the supervisor involved the workers in joint decision making, the more they identified with the company.

Seen as a leader; likeable and reasonable in expectations. Not a driver, bossy, quick to criticize, or unnecessarily strict.

Gives recognition for good work. Praises sincerely, gives recognition in ratings and reports, tells superiors, and recommends promotions.

TRAITS OF TOP MANAGERS (Osburn and Schneeberg 1983)

Have motivation
Set goals
Stress strengths
Emphasize productivity
Concentrate on one area at a time
Take calculated risks
Keep in touch with buyers and suppliers

REFERENCES AND RECOMMENDED READINGS

Hammaker, P. M. and L. T. Rader. 1977. Plain Talk to Young Executives. Homewood, IL: Richard D. Irwin, Inc.

Likert, R. 1967. The Human Organization. New York: McGraw-Hill Book Company.

Osburn, D. D. and K. C. Schneeberger. 1983. Modern Aquacultural Management (Second Edition). Reston, VA: Reston Publishing Company, Inc.

Appendix II

Suggested Steps for Learning to Read People

McCormack (1984) listed seven steps to learning to read people.

1. Listen aggressively. Listen not only to what is said but also to how it is said. People often tell more than they mean to, and if the listener keeps pausing, the slightly uncomfortable silence will make them say even more.
2. Observe aggressively. You can easily learn to interpret certain motions and gestures or to "hear" the statement made by the way a person is dressed.
3. Talk less. You will automatically learn more, hear more, see more, and make fewer blunders. Everyone can and almost everyone should talk less. Ask questions, and don't answer them yourself.
4. Look a second time at a first impression. Scrutinize before you accept the impression as a tenet of a relationship.
5. Take time to use what you've learned. Just before making a phone call or a presentation, stop and think about the reaction you want. From your insight into the other person, what can you do to get this response?
6. Be discreet. Do not tell the other person how insecure you think he or she is. Also, the surest way to expose your insecurity quotient is to talk about your accomplishments. Let others learn of your qualities and achievements elsewhere.
7. Be detached. Force yourself to step back and observe. Then act, using what you've learned, rather than react. If you don't react, you can't overreact.

REFERENCE AND RECOMMENDED READING

McCormack, M. H. 1984. What They Don't Teach You at Harvard Business School. 256 pp. New York: Bantam Books, Inc.

Appendix III

Examples of Enterprise Budgets, Cash Flow, and Credit Repayment Schedules

Assumptions for Catfish Production Budgets and Cash
   Flow Statements
Catfish: 10 Acres, Year 1
Catfish: 10 Acres, Year 2

Source: Pomeroy, R. S., D. B. Luke, and T. Schwedler. 1986. Budgets and cash flow statements for South Carolina Catfish Production. 61pp. Clemson University Agriculture and Rural Sociology. Working Paper: WP052686. Clemson, SC.

ASSUMPTIONS FOR CATFISH PRODUCTION BUDGETS AND CASH FLOW STATEMENTS

1. Land is owned. The land rent charge represents an opportunity cost.
2. Ponds and a water supply (as specified) are in place.
3. Six-inch fingerlings are purchased at $0.15 each. Fingerlings should be purchased only through a reputable local fingerling producer or dealer.
4. Extruded, floating catfish feed, 32% protein, at a cost of $325/ton, is used.
5. Chemicals are applied as recommended at an average cost of $25 per acre for the five stocking rate systems of 1,000 fingerlings per acre and at $35 per acre for the stocking rate systems of 3,500 fingerlings per acre.
6. Tax and insurance are based on a cost of $6 per acre.
7. Repair and maintenance of ponds and equipment are based on a cost of $10 per acre.
8. Two stocking rates for fingerlings are used for the analysis: 1,000/acre and 3,500/acre. The 1,000/acre stocking rate represents an extensive system, and the 3,500/acre rate represents an intensive system.
9. The mortality rate is assumed to be 6%.

10. Feed consumption is based on calculations developed by using the Mississippi State University Catfish Growth Simulation Model (Agricultural Economics Technical Pub. No. 42, January 1983) and on work by Dr. Thomas Lovell of Auburn University (Effect of size on feeding responses of catfish in ponds, *Aquaculture Magazine*, March/April 1984).

11. Each budget has been prepared for two years. The first year begins in March with the stocking of fingerlings. One-third of the catfish are harvested in September and immediately restocked at a rate of one fingerling per fish harvested. The remaining mature fish are harvested in November and immediately restocked at the same rate. The catfish are allowed to grow out through the second year, when one-third are harvested in July and restocked; the rest of the catfish of the second stocking are harvested in September and restocked. The harvest/restocking system is continued. Catfish may be harvested at any time when the desired weight is reached and then restocked immediately with fingerlings.

12. Miscellaneous equipment includes waders, dipnets, scale, baskets, nylon rope, and holding cages.

13. Aeration is not necessary for ponds stocked at 1,000 fingerlings per acre, but is necessary for the ponds stocked at 3,500 fingerlings per acre. Aerators, at 1 hsp per acre, have the following costs (January 1988):

POND SIZE (ACRES)	COST PER UNIT ($)
1	400
2.5	700
5	1,900
10	2,600
20	3,200

Two aerators are needed for the 20 acre pond.

14. Labor:
   a. Preharvest—family (hours): no charge

	1 ACR	2.5 ACR	5 ACR	10 ACR	20 ACR
1,000 stocking rate	125	130	135	170	190
3,500 stocking rate	125	130	150	180	250

   b. Harvest—family (hours): no charge

	1 ACR	2.5 ACR	5 ACR	10 ACR	20 ACR
1,000 stocking rate	60	65	70	100	150
3,500 stocking rate	65	65	70	100	190

c. Harvest—hired (hours): $4.15 hour wage rate

	1 ACR	2.5 ACR	5 ACR	10 ACR	20 ACR
1,000 stocking rate	—	—	—	20	50
3,500 stocking rate	—	—	—	40	80

Department of Agricultural Economics and Rural Sociology

CLEMSON UNIVERSITY
Cooperative Extension Service
Aquaculture Budget

```
CATFISH FOR HUMAN CONSUMPTION
1986 ESTIMATED COST AND RETURNS PER ACRE, YEAR-ROUND MULTIPLE HARVEST SYSTEM
BASED ON RECOMMENDED PRODUCTION PRACTICES FOR COMMERCIAL AQUACULTURE PROCEDURES IN SOUTH CAROLINA
10 ACRE EXISTING POND, 1000 FINGERLINGS PER ACRE STOCKING RATE, FIRST YEAR
```

	UNIT	PRICE OR COST/UNIT	QUANTITY	TOTAL COST-VALUE OF POND	VALUE OR COST PER LB.	VALUE OR COST PER ACRE	YOUR ESTIMATE
1. GROSS RECEIPTS FROM PRODUCTION							
CATFISH	LBS.	$0.75	12850.00	$9637.50			
TOTAL				$9637.50	$0.75	$963.75	_____
2. VARIABLE COSTS							
PREHARVEST							
FINGERLINGS (6 INCH)	EACH	0.150	21200.00	$3180.00	$0.25	$318.00	_____
FEED, 32% PROTEIN	TON	325.000	9.90	$3217.50	$0.25	$321.75	_____
CHEMICALS	ACRE	25.000	10.00	$ 250.00	$0.02	$ 25.00	_____
INSURANCE & TAXES	ACRE	6.000	10.00	$ 60.00	$0.00	$ 6.00	_____
POND REPAIR & MAINT.	ACRE	10.000	10.00	$ 100.00	$0.01	$ 10.00	_____
FAMILY LABOR	HRS.	0.000	170.00	$ 0.00	$0.00	$ 0.00	_____
MACHINE & TRACTOR	ACRE	12.000	10.00	$ 120.00	$0.01	$ 12.00	_____
INTEREST ON OP. CAP.	DOL.	0.120	6807.50	$ 544.60	$0.04	$ 54.46	_____
SUBTOTAL, PREHARVEST:				$7472.10	$0.58	$747.21	_____
HARVEST COSTS							
MACHINE COSTS	ACRE	1.500	10.00	$ 15.00	$0.00	$ 1.50	_____
HIRED LABOR	HOUR	4.150	20.00	$ 83.00	$0.01	$ 8.30	_____
FAMILY LABOR	HOUR	0.000	100.00	$ 0.00	$0.00	$ 0.00	_____
SUBTOTAL HARVEST:				$ 98.00	$0.01	$ 9.80	_____
TOTAL VARIABLE COSTS:				$7570.10	$0.59	$757.01	_____
3. INCOME ABOVE VARIABLE COSTS:				$2067.40	$0.16	$206.74	_____
4. FIXED COSTS:							
MACHINERY, TRACTORS	TOTAL	150.000	1.00	$ 150.00	$0.01	$ 15.00	_____
POND & EQUIPMENT	TOTAL	705.500	1.00	$ 705.50	$0.05	$ 70.55	_____
TOTAL FIXED COSTS				$ 855.50	$0.07	$ 85.55	_____
5. TOTAL OF ABOVE COSTS:				$8425.60	$0.66	$842.56	_____
6. RETURNS TO LAND, MGT, RISK:				$1211.90	$0.09	$121.19	_____
7. OTHER COSTS:							
GEN FARM OVHD	% VAR	8.0%	7570.10	$ 605.61	$0.05	$ 60.56	_____
LAND COST	$/ACRE	$20.00	10.00	$ 200.00	$0.02	$ 20.00	_____
TOTAL OTHER COSTS:				$ 805.61	$0.06	$ 80.56	_____
8. TOTAL COSTS:				$9231.21	$0.72	$923.12	_____
9. RETURNS TO MANAGEMENT, RISK AND FAMILY LABOR:				$ 406.29	$0.03	$ 40.63	_____
BREAKEVEN PRICE, CASH					$0.59		_____
BREAKEVEN PRICE, ALL					$0.72		_____

```
NET RETURNS ABOVE VARIABLE COSTS AT DIFFERENT YIELDS AND PRICES FOR 10 ACRE POND OPERATION.
```

			PRICE PER POUND		
TOTAL YIELD	$0.55	$0.65	$0.75	$0.85	$0.95
12750	-558	717	1992	3267	4542
12800	-530	750	2030	3310	4590
12850	-503	782	2067	3352	4637
12900	-475	815	2105	3395	4685
12950	-448	847	2142	3437	4732

```
NOTE:  ASSUMES CONSTANT VARIABLE COSTS.
```

Source: Clemson University, Agriculture and Rural Sociology. Working Paper: WP052685 by Robert Pomeroy, Dawson B. Luke, and Thomas Schwedler, June 1986. Reproduced with permission.

Department of Agricultural Economics and Rural Sociology

CLEMSON UNIVERSITY
Cooperative Extension Service
Aquaculture Budget

CATFISH FOR HUMAN CONSUMPTION
1986 ESTIMATED COST AND RETURNS PER ACRE, YEAR-ROUND MULTIPLE HARVEST SYSTEM
BASED ON RECOMMENDED PRODUCTION PRACTICES FOR COMMERCIAL AQUACULTURE PROCEDURES IN SOUTH CAROLINA
10 ACRE EXISTING POND, 1000 FINGERLINGS PER ACRE STOCKING RATE, SECOND YEAR

	UNIT	PRICE OR COST/UNIT	QUANTITY	TOTAL COST-VALUE OF POND	VALUE OR COST PER LB.	VALUE OR COST PER ACRE	YOUR ESTIMATES
1. GROSS RECEIPTS FROM PRODUCTION							
CATFISH	LBS.	$0.75	10000.0	$7500.00			
TOTAL				$7500.00	$0.75	$750.00	_____
2. VARIABLE COSTS							
PREHARVEST							
FINGERLINGS (6 INCH)	EACH	0.150	10600.00	$1590.00	$0.16	$159.00	_____
FEED, 32% PROTEIN	TON	325.000	8.30	$2697.50	$0.27	$269.75	_____
CHEMICALS	ACRE	25.000	10.00	$ 250.00	$0.03	$ 25.00	_____
INSURANCE & TAXES	ACRE	6.000	10.00	$ 60.00	$0.01	$ 6.00	_____
POND REPAIR & MAINT.	ACRE	10.000	10.00	$ 100.00	$0.01	$ 10.00	_____
FAMILY LABOR	HRS.	0.000	170.00	$ 0.00	$0.00	$ 0.00	_____
MACHINE & TRACTOR	ACRE	12.000	10.00	$ 120.00	$0.01	$ 12.00	_____
INTEREST ON OP. CAP.	DOL.	0.120	4697.50	$ 375.80	$0.04	$ 37.58	_____
SUBTOTAL, PREHARVEST:				$5193.30	$0.52	$519.33	_____
HARVEST COSTS							
MACHINE COSTS	ACRE	1.500	10.00	$ 15.00	$0.00	$ 1.50	_____
HIRED LABOR	HOUR	4.150	20.00	$ 83.00	$0.01	$ 8.30	_____
FAMILY LABOR	HOUR	0.000	100.00	$ 0.00	$0.00	$ 0.00	_____
SUBTOTAL HARVEST:				$ 98.00	$0.01	$ 9.80	_____
TOTAL VARIABLE COSTS:				$5291.30	$0.53	$529.13	_____
3. INCOME ABOVE VARIABLE COSTS:				$2208.70	$0.22	$220.87	_____
4. FIXED COSTS:							
MACHINERY, TRACTORS	TOTAL	150.000	1.00	$ 150.00	$0.02	$ 15.00	_____
POND & EQUIPMENT	TOTAL	705.500	1.00	$ 705.50	$0.07	$ 70.55	_____
TOTAL FIXED COSTS				$ 855.50	$0.09	$ 85.55	_____
5. TOTAL OF ABOVE COSTS:				$6146.80	$0.61	$641.68	_____
6. RETURNS TO LAND, MGT, RISK:				$1353.20	$0.14	$135.32	_____
7. OTHER COSTS:							
GEN FARM OVHD	% VAR	8.0%	5291.30	$ 423.30	$0.04	$ 42.33	_____
LAND COST	$/ACRE	$20.00	10.00	$ 200.00	$0.02	$ 20.00	_____
TOTAL OTHER COSTS:				$ 623.30	$0.06	$ 62.33	_____
8. TOTAL COSTS:				$6770.10	$0.68	$677.01	_____
9. RETURNS TO MANAGEMENT, RISK AND FAMILY LABOR:				$ 729.90	$0.07	$ 72.99	_____
BREAKEVEN PRICE, CASH					$0.53		_____
BREAKEVEN PRICE, ALL					$0.68		_____

NET RETURNS ABOVE VARIABLE COSTS AT DIFFERENT YIELDS AND PRICES FOR 10 ACRE POND OPERATION.

TOTAL YIELD	$0.55	$0.65	PRICE PER POUND $0.75	$0.85	$0.95
9900	154	1144	2134	3124	4114
9950	181	1176	2171	3166	4161
10000	209	1209	2209	3209	4209
10050	236	1241	2246	3251	4256
10100	264	1274	2284	3294	4304

NOTE: ASSUMES CONSTANT VARIABLE COSTS.

Source: Clemson University, Agriculture and Rural Sociology. Working Paper: WP052685 by Robert Pomeroy, Dawson B. Luke, and Thomas Schwedler, June 1986. Reproduced with permission.

Summary of pond and related equipment fixed costs: 10-acre pond, 1,000 fingerlings stocking rate

SYSTEM EQUIPMENT	YEARS OF LIFE	NEW INVEST.	AVERAGE INVEST.	ANNUAL DEPREC.	INTEREST (12%)	ANNUAL TAX AND INSUR.	TOTAL FIXED COST
Harvest seines	5	700	350	70	42	3	115
Miscellaneous	5	450	225	45	27	2	74
Oxygen and chemical kit	5	150	75	15	9	1	25
Transport tank	10	1,000	500	50	60	4	114
Boat and motor	10	600	300	30	36	2	68
Storage shed	20	3,500	1,750	88	210	13	311
Total		6,400	3,200	298	384	24	706

CATFISH: 10 ACRES, YEAR 1

Credit Repayment Schedule for Current Debts, 1986

	OWED	INTEREST	PRINCIPAL	BALANCE	YRS. TO PAY	INT. RATE
	32,730.00	3,927.60	0.0	32,730.00	10	12%
	6,400.00	768.00	0.0	6,400.00	5	12%
Totals	39,130.00	4,695.60	0.0	39,130.00		

Note: The loan of $32,730 for ten years is for pond construction, pump system, storage shed, and feed dispenser.
The loan for $6,400 for five years is for other required equipment and supplies. Only interest is paid during the first year. Interest and principal payments begin in July of the second year. An operating loan that is due at the end of the first year is available at 12%.

Financial Statement, 1986

| | ASSETS | |
	BEGINNING	END
Cash on hand	0	19,418
Pond	27,230	15,000
Stor. shed	3,500	3,325
Feed desp.	2,000	1,868
Boat	600	540
Trans. tank	2,000	1,800
Aeration eq.	2,500	2,250
Misc.	1,300	1,040
Subtotal	39,130	45,241
Total assets	39,130	45,241

| | LIABILITIES | |
	BEGINNING	END
Short-term debt	0	0
Intermediate debt	6,400	6,400
Long-term debt	32,730	32,730
New debt		13,710
Interest on new debt		1,094
Total debt	39,130	53,935
Net worth	0	3,694

Catfish: 10 Acres, Year 1: Cash Flow Projections for 1986

KIND	JAN	FEB	MAR	APR	MAY	JUNE	JULY	AUG	SEP	OCT	NOV	DEC	YEAR
Beg. bank balance	0												
Income Items													
Catfish	0	0	0	0	0	0	0	0	11,138	0	22,613	0	33,750
Totals	0	0	0	0	0	0	0	0	11,138	0	22,613	0	33,750
Cost Items													
Interest	0	0	0	0	0	0	0	0	0	0	0	4,696	4,696
Fingerlings	0	0	5,565	0	0	0	0	0	1,853	0	3,712	0	11,130
Feed	0	0	382	617	869	1,202	1,934	2,592	1,357	1,852	382	57	11,245
Chemicals	0	0	35	35	35	35	35	35	35	35	35	35	350
Aeration	0	0	0	0	32	32	32	32	32	0	0	0	161
Pond maint.	0	0	10	10	10	10	10	10	10	10	10	10	100
Insur. and tax	0	0	0	0	0	0	0	0	0	0	0	60	60
Mach. cost	0	0	14	7	7	7	7	7	36	7	36	7	135
Hired labor	0	0	0	0	0	0	0	0	83	0	83	0	166
Totals	0	0	6,006	670	954	1,287	2,018	2,676	3,406	1,905	4,257	4,865	28,043
Net cash income	0	0	−6,006	−670	−954	−1,287	−2,018	−2,676	7,732	−1,905	18,355	−4,865	5,707
Capital purchases	0	0	0	0	0	0	0	0	0	0	0	0	0
Old debt repaid	0	0	0	0	0	0	0	0	0	0	0	0	0
New money borrowed	100	0	6,006	670	954	1,287	2,018	2,676	0	0	0	0	13,710
Checkbook balance	100	100	100	100	100	100	100	100	7,832	5,927	24,282	19,418	19,418
Int.—new money borrowed	1	1	61	68	77	90	110	137	137	137	137	137	1,094
Accum int.—new money	1	2	63	131	208	298	409	546	683	820	957	1,094	1,094

System description and assumptions: A 10-acre pond for catfish production, year 1. Pond construction begins in January. Loans are available to cover construction costs and to purchase necessary equipment in January. The pond is stocked with 37,100 fingerlings in March. A 6% death loss is assumed. About one-third of the fish are harvested in September (14,850 pounds) and two-thirds in November (30,150 pounds). Restocking takes place in the same months fish are harvested.

First year's operating fund requirement is $13,710 plus $1,094 in interest. The total obligation for the operating loan is $14,804. It is assumed that the operating loan and interest are paid off in December of the first year from the checkbook balance of $19,418. This will pay off the new money borrowed and interest — new money borrowed and leave a checkbook balance to carry forward to year 2 of $4,614 ($19,418 − $14,804 = $4,614) at the end of the year.

CATFISH: 10 ACRES, YEAR 2

Credit Repayment Schedule for Current Debts, 1987

	OWED	INTEREST	PRINCIPAL	BALANCE	YRS TO PAY	INT. RATE
	32,730.00	3,822.19	2,158.11	30,571.88	9	12%
	6,400.00	703.18	1,327.11	5,072.87	4	12%
Totals	39,130.00	4,525.37	3,485.22	35,644.75		

Financial Statement, 1987

	ASSETS	
	BEGINNING	END
Cash on hand	4,614	6,905
Pond	15,000	15,000
Stor. shed	3,325	3,159
Feed desp.	1,868	1,745
Boat	540	486
Trans. tank	1,800	1,620
Aeration eq.	2,250	2,025
Misc.	1,040	832
Subtotal	30,437	31,771
Total assets	30,437	31,771

	LIABILITIES	
	BEGINNING	END
Short-term debt	0	0
Intermediate debt	6,400	5,073
Long-term debt	32,730	30,572
New debt		14
Interest on new debt		1
Total debt	39,130	35,659
Net worth	−8,693	−3,888

REFERENCE AND RECOMMENDED READING

Pomeroy, R. S., D. B. Luke, and T. Schwedler. 1986. Budgets and cash flow statements for South Carolina Catfish Production. Working Paper: WP052686. Clemson, SC. Clemson University Agriculture and Rural Sociology.

Catfish: 10 Acres, Year 2: Cash Flow Projections for 1987

	JAN	FEB	MAR	APR	MAY	JUNE	JULY	AUG	SEP	OCT	NOV	DEC	YEAR
Beg. bank balance	4,614												
Income Items													
Catfish	0	0	0	0	0	0	8,663	0	17,588	0	0	0	26,250
Totals	0	0	0	0	0	0	8,663	0	17,588	0	0	0	26,250
Cost Items													
Interest	0	0	0	0	0	0	776	778	760	749	737	725	4,525
Fingerlings	0	0	0	0	0	0	1,853	0	3,712	0	0	0	5,565
Feed	57	57	574	1,016	1,118	1,385	1,493	1,409	1,082	665	512	57	9,425
Chemicals	29	29	29	29	29	29	29	29	29	29	29	29	350
Aeration	0	0	0	0	32	32	32	32	32	0	0	0	161
Pond Maint.	8	8	8	8	8	8	8	8	8	8	8	8	100
Insur. and Tax	0	0	0	0	0	0	0	0	0	0	0	60	60
Mach. Cost	5	5	5	5	5	5	42	5	42	5	5	5	135
Hired Labor	0	0	0	0	0	0	83	0	83	0	0	0	166
Totals	100	100	617	1,058	1,193	1,460	4,317	2,262	5,748	1,457	1,292	885	20,488
Net Cash Income	-100	-100	-617	-1,058	-1,193	-1,460	-4,346	-2,262	11,839	-1,457	-1,292	-885	5,762
Capital Purchases	0	0	0	0	0	0	0	0	0	0	0	0	0
Old Debt Repaid	0	0	0	0	0	0	-559	-557	-575	-586	-598	-610	-3,485
New Money Borrowed	0	0	0	0	0	14	0	0	0	0	0	0	14
Checkbook Balance	4,514	4,415	3,798	2,739	1,547	100	3,887	1,068	12,332	10,289	8,399	6,905	6,905
Int. New Money Borrowed	0	0	0	0	0	0	0	0	0	0	0	0	0
Accum Int. — New Money	0	0	0	0	0	0	0	0	1	1	1	1	1

Appendix IV

Life-Cycle Costing

Source: U.S. Department of the Treasury, Internal Revenue Service Document 7069 (12-85).

For evaluating most investments of the federal government, the Office of Management and Budget (OMB) has specified that a real rate of 10% must be used. The formula for calculating multipliers is derived from the concept of the time value of money. A dollar today is worth more than a dollar a year from now because today's dollar may be invested and may be worth more than a dollar in one year.

For example, a dollar invested today at 10% interest will be worth $1.10 in one year. To derive a general formula for this calculation, let P = present amount, I = interest rate, and F = future worth. Thus:

$$F = P + (P \times I)$$
$$= \$1.00 + (\$1.00 \times .10)$$
$$= \$1.00 + \$.10$$
$$= \$1.10$$

That same $1.00 invested for two years at 10% will be worth $1.21. That is because the $1.10 available at the end of the first year is, in effect, reinvested again at 10%. So, 10% of $1.10 ($0.11) is added to the $1.10 to give us $1.21 at the end of the two years.

Generally, then:

$$F = P + (P \times I) + I\,[P + (P \times I)]$$
$$= \$1.00 + (\$1.00 \times .10) + .10\,[\$1.00 + (\$1.00 \times .10)]$$
$$= \$1.10 + .10\,[\$1.00 + .10]$$
$$= \$1.10 + \$.11$$
$$= \$1.21$$

161

This general formula is reiterated for each year the money collects interest, thereby allowing us to determine a formula for any year, which is $F_n = P(1 + I)$, where N = number of years. If we wanted to determine the present amount, P, rather than the future worth, F, algebra reveals that:

$$P = F_n \times \frac{1}{(1 + I)}^n$$

For convenience, a list of present values using the 10% discount factor is as follows:

End-of-Year vs. Average Discount Factors (10%)

YEARS	END-OF-YEAR FACTOR	AVERAGE FACTOR
0	1.000	
1	0.909	0.954
2	0.826	0.867
3	0.751	0.788
4	0.683	0.717
5	0.621	0.652

The factors depicted in the center column above are termed *end-of-year* factors. This label is appropriate because they are derived under the assumption that cash flows occur precisely at the ends of years. In the real world, this is generally not the case. Costs are usually dispersed throughout the year. Thus, a more realistic discount factor would be one that is an average for the year. The table column on the right illustrates the conversion from end-of-year to average factors.

If costs are uniform over several years, you can multiply the annual cost by the cumulative discount factor (Table 11-3 of Chapter 11) for those years, rather than multiplying each year's costs by the yearly discount factor.

Appendix V

Sample Calculation to Determine the Number of Fish That Can Be Reared in a Unit (Tank) That Receives 5 Gallons Per Minute (GPM) Water Flow

---

Primary Feed Level Calculation
Alternate Feed Level Calculation

Procedure: Set minimum exchange rate; determine available DO; determine carrying capacity as kg of feed; determine number and weight of fish that can be reared on fixed feed level.

Assumptions: Exchange rate of 3 per hour and unit volume of 13.4 ft^3 or 370 L

$$\text{flow} = 3/h \times 379\ L = 1137\ L/h$$
$$= 19\ LPM\ (5\ gpm)$$

PRIMARY FEED LEVEL CALCULATION

To achieve loading equivalent to one complete use, reduce DO from 90% of saturation (minimum inflow concentration) to 6.5 mg/L O_2 at 8.9°C (48°F):

$$\text{Saturation DO} = 11.6\ mg/L$$
$$90\%\ \text{saturation} = 10.4\ mg/L$$
$$\text{Available } O_2 = 10.4 - 6.5$$
$$= 3.9\ mg/L$$

163

$$24\text{-hour available } O_2 = 3.9 \text{ mg/L} \times 1140 \text{ LPH} \times 24 \text{ h}$$

$$= 106{,}704 \text{ mg } O_2$$

$$= 107 \text{ g } O_2$$

$$O_2 \text{ requirement for salmonids} = 200 \text{ g/kg feed}$$

$$\text{feed per day} = 107/_{200}$$

$$= 0.53 \text{ kg feed/day/unit}$$

ALTERNATE FEED LEVEL CALCULATION

Westers (Michigan Manual, 1981 revision) shows that 35 LPM flow is required per kilogram of feed

$$\text{LPM/kg} = \frac{200}{1.44 \times (10.4 - 6.5)} = 35.6 \text{ LPM}$$

then:

$$\frac{19 \text{ LPM}}{35.6 \text{ LPM/kg feed}} = 0.53 \text{ kg}$$

Alter the calculations to conform to hatchery feeding schedules; for instance, many hatcheries feed during the work day and 70% of the 24-hour oxygen demand occurs during the 12-hour feed effect day:
(Primary calculation)
Then:

$$12\text{-hour available } O_2 = 3.9 \times 1140 \times 12$$

$$= 53.5 \text{ g } O_2$$

$$O_2 \text{ required} = 200 \text{ g/kg feed} \times 0.7$$

$$= 140$$

$$\text{Feed/day} = 53.5/140 = 0.38 \text{ kg feed}$$

(Alternate calculation)

$$O_2 \text{ required} = \frac{200}{(1.44 \times 0.07) \times 3.9} = 50.9 \text{ LPM/kg feed}$$

$$19 \text{ LPM}/50.9 \text{ LPM/kg feed} = 0.37 \text{ kg feed}$$

To determine the number of fish, use a feeding schedule (as planned for use in production) to find the percent body weight to feed. Divide the weight of the food by the percent body weight to find the weight of fish limit in the unit. Use conversion charts (Piper et al. 1981) to estimate number of fish. The following table gives an example for fish 5–6.5 inches long.

Table of Fish Requirement

| LENGTH | | FEED | WEIGHT/UNIT | | NUMBER | POUNDS |
CM	IN.	% BW	KG	LB	PER UNIT	PER FT^3
12.7	5.0	1.5	24.7	54.4	1740	4.06
14.0	5.5	1.4	26.4	58.3	1399	4.35
15.2	6.0	1.3	28.5	62.8	1162	4.69
16.5	6.5	1.2	30.8	68.0	993	5.07

REFERENCES AND RECOMMENDED READINGS

Piper, R. G., I. B. McElwain, L. E. Orme, J. P. McCraren, L. G. Fowler, and J. R. Leonard. 1982. Fish Hatchery Management. Washington, DC: U.S. Fish and Wildlife Service.
Westers, H. 1984. Principles of Intensive Culture. Lansing, MI: Michigan Dept. of Natural Resources.

Appendix VI

Sample Problems on Production Capacity Assessment (PCA)

PROBLEMS

1. One places 300 g of fish into each of five jars plumbed in serial reuse; water is gravity aerated to roughly 90% of saturation between each unit; the weight of the fish in the first jar doubles in 30 days, whereas the weight increase in each successive jar is only half that of the one above it; the group of fish in each jar remove roughly an equal amount of oxygen, 1.5 mg/L. What, then, is the $ECOC_{50}$ (the point of 50% growth reduction in terms of cumulative oxygen demand)?

2. Same as question 1, but fish have an O_2 consumption rate of 2.5 mg DO/L times X, where X = kilograms per liter of fish in the unit.

3. Same as question 1, but the weight increase in each jar after the first is roughly 90% of that in the preceding jar.

4. What is the number of serial rearing units that should not be exceeded to ensure that all fish will have a growth rate of more than half the maximum obtainable, as described in problem

   1: _____

   2: _____

   3: _____

5. In a hatchery start-up study, and while construction was going on, the manager placed the first completed series of six raceways into production. Not only were other raceways still under construction, but even more (additional) raceways were planned for a future phase of construction. The manager stocked fish at a density he considered normal, which turned out to be 500 rounds per raceway. However, even though he used aerators between raceways, the manager could not resist using a greater than planned amount of the abundant water, and actual loading (DO consumption/flow) was only one-third of that planned for the station at full production. In the first 30 days, the manager found the following production results:

UNIT	WEIGHT GAIN	O_2. CONSP.
1	150	0.8
2	155	0.9
3	136	0.8
4	120	0.8
5	105	0.7
6	81	0.6

In a review meeting with the manager's supervisor and two other hatchery managers, one of the managers suggested the addition of two more raceways in each series to take advantage of the obvious production potential. However, the supervisor, who was primarily an administrator and not familiar with fish culture, was cautious. In fact, the supervisor hoped to reduce the planned number of raceways to lower construction costs, and he asked each of the hatchery managers to take a copy of the data and to send him a written recommendation within a week. Based on the PCA concept, what is the maximum number of raceways in each series that should be used (according to these production data) to ensure that at least 50% of maximum growth can be effected in all raceways? Assume that oxygen consumption would triple under full production.

Answers:

1.

REARING UNIT	WEIGHT (G) INCREASE	SG	CUMULATIVE O_2 CONSUMPTION
1	300	1.00	1.5
2	150	0.59	3.0
3	75	0.32	4.5
4	37.5	0.17	6.0
5	18.75	0.09	7.5

SGR of $1.00 \times 50\% = 0.50$
$ECOC_{50} = 4.06$

2.

REARING UNIT	TOTAL WEIGHT	DO CONSP.	CUMULATIVE DO CONSUMPTION
1	0.600	1.500	1.500
2	0.450	1.125	2.625
3	0.375	0.938	3.563
4	0.338	0.845	4.408
5	0.319	0.798	5.206

$ECOC_{50} = 3.19$

3.

REARING UNIT	WEIGHT INCREASE	SG
1	300	1.00
2	270	0.93
3	243	0.86
4	219	0.79
5	197	0.73

50% of 1.00 = 0.50
$ECOC_{50}$ = 12.5

4. The number of rearing units that will have an oxygen consumption approaching, but not exceeding, the $ECOC_{50}$, are:

1: _____ 2 _____
2: _____ 2 _____
3: _____ 8 _____   (12.5/1.5)

5.

O_2 CONSUMPTION	CUMULATIVE O_2 CONSUMPTION	SG
.8	0.8	.38
.9	1.7	.39
.8	2.5	.35
.8	3.3	.31
.7	4.0	.28
.6	4.6	.22

SG_{50} = .19
$ECOC_{50}$ = 5.9
(y intercept = 0.44)
(slope = -0.043)

$3 \times$ mean O_2 consumption = 2.4
and $ECOC_{50}$/mean O_2 consumption = number of uses
or
5.9/2.4 = 2.45
Therefore the manager can use only two raceways in full production. The facility is grossly overdesigned.

Index